Lecture Notes in Computer Science 9420

Commenced Publication in 1973
Founding and Former Series Editors:
Gerhard Goos, Juris Hartmanis, and Jan van Leeuwen

More information about this series at http://www.springer.com/series/8851

Ngoc Thanh Nguyen · Ryszard Kowalczyk
Béatrice Duval · Jaap van den Herik
Stephane Loiseau · Joaquim Filipe (Eds.)

Transactions on Computational Collective Intelligence XX

Springer

Editors-in-Chief
Ngoc Thanh Nguyen
Faculty of Computer Science
    and Management
Wroclaw University of Technology
Poland

Ryszard Kowalczyk
Swinburne University of Technology
Hawthorn
Australia

Guest Editors
Béatrice Duval
LERIA - UFR Sciences
Angers Cedex 01
France

Stephane Loiseau
LERIA - UFR Sciences
Angers Cedex 01
France

Jaap van den Herik
Leiden University
Leiden
The Netherlands

Joaquim Filipe
INSTICC
Polytechnic Institute of Setubal
Setubal
Portugal

ISSN 0302-9743   ISSN 1611-3349   (electronic)
Lecture Notes in Computer Science
ISBN 978-3-319-27542-0   ISBN 978-3-319-27543-7   (eBook)
DOI 10.1007/978-3-319-27543-7

Library of Congress Control Number: 2015945327

Printed on acid-free paper

This Springer imprint is published by SpringerNature
The registered company is Springer-Verlag GmbH Berlin Heidelberg

Transactions on Computational Collective Intelligence XX

Preface

The present special issue of the journal *Transactions on Computational Collective Intelligence (TCCI)* includes extended and revised versions of a set of selected papers from the 2014 and 2015 editions of the International Conference on Agents and Artificial Intelligence – ICAART.

The interdisciplinary fields of agents and artificial intelligence (AI) involve a large number of researchers who devote themselves to the study of theoretical and practical issues related to areas such as multi-agent systems, software platforms, agile management, distributed problem solving, distributed AI in general, knowledge representation, planning, learning, scheduling, perception, data mining, data science, reactive AI systems, evolutionary computing, and other topics related to intelligent systems and computational intelligence.

This special issue presents 11 research papers with novel concepts and applications in the aforementioned areas.

The first six papers are revised and extended versions of papers presented at ICAART 2014, including four papers related to the topic "Intelligent Agents," involving either (a) multi-agent systems, cooperation, and coordination, as in the case of "Abstraction of Heterogeneous Supplier Models in Hierarchical Resource Allocation" (by Alexander Schiendorfer, Gerrit Anders, Jan-Philipp Steghöfer, and Wolfgang Reif) and "Extensibility-Based Multiagent Planner with Plan Diversity Metrics" (by Jan Tožička, Jan Jakubův, Karel Durkota, and Antonín Komenda) or (b) research work more focused on agent models that enable some degree of interaction or context understanding, as in the case of "Developing Embodied Agents for Education Applications with Accurate Synchronization of Gesture and Speech" (by Jianfeng Xu, Yuki Nagai, Shinya Takayama, and Shigeyuki Sakazawa) and "Shape Recognition Through Tactile Contour Tracing – A Simulation Study" (by André Frank Krause, Nalin Harischandra, and Volker Dürr). From the set of ICAART 2014 best papers, we also selected the following two, which presented research work related to "Machine Learning and Pattern Recognition," an important area of AI, highlighting new developments in the unsolved problems of feature selection, by using filter methods and support vector machine, and multi-objective optimization, by using knowledge gradient and upper confidence bound policies, namely: "Real-Time Tear Film Classification Through Cost-Based Feature Selection" (by Verónica Bolón-Canedo, Beatriz Remeseiro, Noelia Sánchez-Maroño, and Amparo Alonso-Betanzos) and "Scalarized and Pareto Knowledge Gradient for Multi-objective Multi-armed Bandits" (by Saba Yahyaa, Madalina M. Drugan, and Bernard Manderick).

The other five papers are revised and extended versions of papers presented at ICAART 2015, including three papers related to the topic "Intelligent Agents," involving some form of multi-agent system cooperation and coordination, either in a simulation environment, optimizing some sort of behavior, or reasoning with other agents interactively. This set of papers includes: "Concurrent and Distributed Shortest-Path Searches in Multiagent-Based Transport Systems" (by Max Gath, Otthein Herzog, and Maximilian Vaske), "SAJaS: Enabling JADE-Based Simulations" (by Henrique Lopes Cardoso), "Strategic Negotiation and Trust in Diplomacy – The DipBlue Approach" (by André Ferreira, Henrique Lopes Cardoso, and Luis Paulo Reis). A fourth paper was selected that, in spite of its focus on multi-agent systems, used a different "reactive-AI" approach, namely: "Overcoming Limited Onboard Sensing in Swarm Robotics Through Local Communication" (by Tiago Rodrigues, Miguel Duarte, Margarida Figueiró, Vasco Costa, Sancho Moura Oliveira, and Anders Lyhne Christensen). Finally, we selected a paper related to machine learning in AI, namely: "A Question of Balance: The Benefits of Pattern Recognition when Solving Problems in a Complex Domain" (Martyn Lloyd-Kelly, Fernand Gobet, and Peter C.R. Lane).

We believe that all the papers in this special issue will serve as a reference for students, researchers, engineers, and practitioners who conduct research in the areas of agents and AI. We hope that the readers will find new inspiration for their research and may join the ICAART community in the future.

We would like to thank all the authors for their contributions and all the reviewers who helped ensure the quality of this publication. Finally, we would also like to express our gratitude to the LNCS editorial staff of Springer, in particular Prof. Ryszard Kowalczyk for all his patience and availability during this process.

November 2015

Béatrice Duval
Jaap van den Herik
Stephane Loiseau
Joaquim Filipe

Transactions on Computational Collective Intelligence

This Springer journal focuses on research in applications of the computer-based methods of computational collective intelligence (CCI) and their applications in a wide range of fields such as the Semantic Web, social networks, and multi-agent systems. It aims to provide a forum for the presentation of scientific research and technological achievements accomplished by the international community.

The topics addressed by this journal include all solutions of real-life problems for which it is necessary to use computational collective intelligence technologies to achieve effective results. The emphasis of the papers published is on novel and original research and technological advancements. Special features on specific topics are welcome.

Editor-in-Chief

Contents

Developing Embodied Agents for Education Applications with Accurate Synchronization of Gesture and Speech

Jianfeng Xu[✉], Yuki Nagai, Shinya Takayama, and Shigeyuki Sakazawa

Smart Home and Robot Application Laboratory, KDDI R&D Laboratories, Inc.,
2-1-15 Ohara, Fujimino-shi, Saitama 356-8502, Japan
{ji-xu,yk-nagai,sh-takayama,sakazawa}@kddilabs.jp

Abstract. Embodied agents have great potential for education field, which are promising to maximize the learner's learning gains and enjoyment. In many education applications, multimodal representation of embodied agents is a powerful approach for obtaining the above benefit, which requires accurate synchronization of gesture and speech. For this purpose, we investigate the important issues in synchronization as a practical guideline for our algorithm design through a precedent case study and propose a two-step synchronization method. Our case study reveals that two issues (i.e. duration and timing) play an important role in synchronizing of gesture with speech. Considering the synchronization problem as a motion synthesis problem instead of a behavior scheduling problem used in the conventional methods, we employ a motion graph technique with constraints on gesture structure for coarse synchronization in a first step and refine this further by shifting and scaling the gesture in a second step. Subjective evaluation has demonstrated that the proposed method achieves more accurate synchronization with respect to both duration and timing, and higher motion quality than the state-of-the-art methods.

Furthermore, we have implemented the proposed synchronization method in an authoring tool for education applications. We have conducted several experiments in a university, whose results have demonstrated that our system makes the creation of attractive animations easier and faster (only about 10 % operation time) than manual creation of equal quality, and it is effective to use embodied agents in education applications.

Keywords: Embodied agents · Education applications · Multimodal synchronization · Gesture · Motion graphs · Dynamic programming

1 Introduction

Over the past two decades, many embodied agents have been developed for various fields such as the interface agents in human computer interaction (e.g. Rea [5], Greta [13,29], and RealActor [42]), virtual training environments, portable personal navigation guides, interactive storytelling systems, interactive online

© Springer-Verlag Berlin Heidelberg 2015
N.T. Nguyen et al. (Eds.): TCCI XX, LNCS 9420, pp. 1–22, 2015.
DOI: 10.1007/978-3-319-27543-7_1

characters, and automated presenters/commentators [12,19,20,31–33]. For instance, a virtual human presenter is designed for weather forecasting, and slide presentation in the work of Noma et al. [31]. It is reported that a navigation agent such as a guide in a theater building [19] or a university campus [33] is helpful to visitors. As one of popular fields for embodied agents, this paper will focus on education applications [40]. The contributions of this paper include not only the technical contributions for embodied agents but also the implementation of an authoring tool to easily create the agent animations and effectiveness evaluation in real educational environments.

A typical education application is that a student in a school or an employee in a company watches a video where a lecturer gives a talk using lecture slides or blackboard. In this paper, an embodied agent will serve as the lecturer in the video (see an example in Fig. 9), which is expected to maximize the learner's learning gains and enjoyment. It is believed that multimodal representation of embodied agents is essential for obtaining the above benefit [7,30], which requires accurate synchronization of gesture and speech [25]. Basically, animators spend an enormous amount of effort using either intuition or motion capture to achieve accurate synchronization. Although synchronization description schemes (e.g. the Behavior Markup Language, BML [15]) have been proposed and widely used in the academic field [21], it remains a challenge to produce such synchronization automatically. In this paper, we propose an effective synchronization method using a different philosophy than the conventional methods. The basic idea is to consider the synchronization problem as a motion synthesis problem instead of a behavior scheduling problem where the gesture motions are re-scheduled in the timeline according to the speech [7,21,27]. The greatest benefit is that we can significantly improve both the synchronization accuracy and motion quality simultaneously.

Psychological research has shown that gesture and speech have a very complex relationship [24]. Although they are believed to share a common thought source, the hypothesized *growth point* [23][1], the relationship between gesture and speech is many-to-many. For example, to emphasize a word in an utterance, one may use a beat gesture, a nod, or an eyebrow. On the other hand, a nod may mean confirmation rather than emphasis. Furthermore, many other factors affect the relationship between gesture and speech such as personality, gender, culture, conversational context, etc. [21,27]. For example, Japanese talk to each other with nodding, but a nod means attentiveness rather than agreement. In addition, human perception is highly sensitive to the synchronization of speech and gesture. Although the temporal tolerance is basically dependent on the content and human subject, it is believed that high accuracy (e.g. 150 ms) is required by most of human subjects [25]. For example, a level of the phoneme is perceptible for most of people to watch a speaker's mouth movements [22].

In the field of embodied conversational agents [6,7] and human-robot interaction [28], the synchronization of gesture and speech is based on a common

---

[1] A growth point is assumed to be a minimal psychological unit with special focus on speech-gesture synchrony and co-expressivity.

practice that synchronizes the gesture stroke (see the definition in Sect. 3) with the accented syllable of the accompanying speech [27], which is also suggested by the *growth point* theory [23]. Based on this, the latest system [21] uses an offset/scaling technique for synchronization of gesture and speech.

Due to the absence of practical guidelines on automating synchronization, in this paper, we investigate the important issues in the manual synchronization of gesture with speech. Two similar but not identical scripts are prepared. We examine the differences among manually created animations of the two scripts and discover practical guidelines for our algorithm. As a result, the above case study reveals that two issues (i.e. duration and timing) play an important role in the manual synchronizing of gesture with speech.

To automatically produce accurate synchronization of gesture and speech, in essence, we consider the synchronization problem to be a motion synthesis problem with certain constraints. At the same time, we observe that many gestures are cyclic or use similar poses, which results in the adoption of the motion graph technique [1,16,17]. The motion graph technique is reported to be a powerful tool for synthesizing natural motion from an original motion with constraints such as motion duration [1,16,17]. In addition, it is well known that gestural motion has a special temporal structure, which is important in the synchronization of gesture and speech [27]. Our experimental results show that the proposed algorithm works well in our scenarios.

In this paper, our technical contributions are summarized as follows.

1. With a case study, we have discovered that two issues (i.e. duration and timing) play an important role in the manual synchronizing of gesture with speech, which becomes a practical guideline for our algorithm.
2. We propose a two-step algorithm based on the motion graph technique [1, 16,17] with a temporal structure of gestures that deals with the issues of duration and timing. In the first step, we synthesize a new motion that is coarsely synchronized with the speech. In the second step, we further refine the synchronization by shifting and scaling the synthesized motion.

In addition, targeted at non-professional content creators (e.g., a teacher in school or a staff in a company), we implement our system as an easy-to-use authoring tool with a user-friendly user interface, which outputs a synthesized animation with facial expressions and gestures synchronized with the audio signal. Our authoring system allows the user to semi-automatically create animation in only about 10 % of the time required for manual creation. For effectiveness evaluation, we create some animations for education applications by our authoring tool and conduct several experiments in a university's class, where we get rather positive results and feedback.

This paper is organized as follows. Section 2 briefly surveys some techniques and systems related to our approach. Section 3 describes the precedent case study and the proposed algorithm for synchronizing gesture with speech in detail. In Sect. 4, we report our experimental results about the synchronization, including a subjective evaluation that compares the proposed method to the conventional method [21]. In Sect. 5, we introduce our authoring tool that can output a rich animation with facial expressions and gestures synchronized with speech. In Sect. 6, we introduce the applications to education using the authoring tool, and report the effectiveness evaluation for education applications. Finally, we present our conclusions and future work in Sect. 7.

2 Related Work

In this section, we briefly survey the embodied agent systems for education applications (Sect. 2.1), synchronization techniques of gesture and speech (Sect. 2.2), and the motion graph techniques used in our system (Sect. 2.3).

2.1 Embodied Agents for Education Applications

Many works have demonstrated that embodied agents with multimodal representation are appealing to users [5,12,19,20,31–33]. For a comprehensive survey of embodied agent systems, we recommend reading Cassell's or Nishida's book [6,30]. Especially, digital education is attracting much attention from both academic and industrial fields with the rapid development of mobile devices like tablet and education contents like MOOC (Massive Open Online Course). Although some embodied agent systems such as [4,14,36] are reported in education applications as follows, few of them have evaluated the effectiveness of embodied agents while this paper will further give the evaluation of agent's effectiveness.

Rist et al. [36] reviewed several embodied presentation agents developed at German Research Centre for Artificial Intelligence (DFKI), which include a single, TV-style presentation agent, dialogue systems, and multiple interactive characters. Beskow et al. [4] designed an interactive talking agent (head only) for language learning. Ieronutti et al. [14] implemented a virtual human for education and training with web technologies (Web3D). Many other systems have been developed as the survey paper [40] mentioned in virtual learning environments.

2.2 Synchronization of Gesture and Speech

Most embodied agent systems are composed of three sequentially executed blocks: audio/text understanding, behavior selection, and behavior editing [7]. Many techniques are available for audio/text understanding [21,41], which provides the needed acoustic and semantic information for behavior selection. Especially, by performing deep analysis of syntactic, semantic and rhetorical structures of the utterance, [21] achieves semantically appropriate behavior, which is their central contribution. For behavior selection, the de-facto method is a rule-based approach [7,21,28] that maps keywords to behaviors or behavior categories by a large set of predefined rules. For behavior editing, existing systems focus mainly on hand trajectory modification by physical simulation [27] or cubic spline interpolation [28]. Unfortunately, there are as yet few techniques for multimodal synchronization, although this is believed to be essential to properly convey the emotional component of communication. Lip synchronization is widely used in embodied agent systems thanks to TTS (text to speech) techniques [8]. For synchronization of gesture and speech, the early work [7] aligns the timing of gesture motions with text words, and the latest paper [21] improves the synchronization level to gesture phases using the offset and scaling approach, in which the timing of the stroke phase in gesture motion is aligned with the speech. However, such an approach will change the quality of motions and even the emotional state of gestures if the scaling factor is too large [44].

On the other hand, psychological research on multimodal synchronization continues and has provided many valuable insights for embodied agent systems. The hypothesized *growth point*, proposed by [23], is a well-known theory to explain the phenomenon of synchronization of gesture and speech. Moreover, based on the fact that the structure of gesture is different from other human motions like dancing, [27] pointed out that the stroke phase should be synchronized with the accented syllable of the accompanying speech. However, these discoveries only specify a result without providing the processing needed to produce it automatically [15].

2.3   Motion Graphs

To re-use motion data in motion synthesis, graph-based representations are demonstrated to be a very powerful tool. Many related studies have been reported, as briefly summarized in [34]. In the game industry, these graphs, called *move trees*, are originally created manually [26]. Several automatic methods are independently proposed to construct motion graphs in SIGGRAPH 2002 [1,16,17]. Basically, a motion graph is a directed graph structure of possible connections between motions, turning the set of initial clips from a list of sequences into a connected graph of movements.

Since then, many variations of motion graphs have been reported. Gleicher et al. [11] present *snap-together graphs* where common poses are used as hubs in the graph. Safonova and Hodgins [38] create an *interpolated motion graph* by combining the standard motion graph and interpolation techniques. Zhao et al. [46] further use interpolation of motion segments of the same contact to add additional data to the database in their *well-connected motion graph*. Besides interpolation, continuous constrained optimization is introduced in the *optimization-based graph* by Ren et al. [35]. Beaudoin et al. [3] cluster similar motions to get an understandable graph structure referred to as a *motion-motif graph*. However, these graph-based representations cannot directly be employed because we need to consider the gesture's temporal structure and define the necessary properties for synchronization such as the motion duration. This paper will extend the conventional motion graph techniques for our task.

Through motion graphs, motion synthesis is cast as a searching problem for a path that minimizes an objective function [10]. The searching strategy highly depends on the application purpose. Given user constraints, Kovar et al. [16] improve the depth-first search to obtain graph walks using a branch-and-bound strategy and incremental search. Lee et al. [17] use the greedy best-first search approach and traverse only a fixed number of frames to maintain a constant rate of motion. Arikan et al. [1] develop a hierarchical, randomized search strategy. They present another method [2] that uses a dynamic programming approach and coarse-to-fine refinement to search for the motion sequence. Dynamic programming, the complexity of which is linear to the path length, is also adopted by Lee and Lee [18].

3   Proposed Synchronization Algorithm

In state-of-the-art systems [21,27], the number of gestures in the database is not very large, amounting to just dozens of available gestures, which is comparable to the number used by a TV talk show host [27]. However, a human being performs each gesture variably according to the context, e.g. synchronizing the gesture with speech. Therefore, gesture variation is rather large in human communication. This indicates that the task of gesture synchronization is in essence to synthesize a new motion from a generic one for a particular portion of speech, which is a motion synthesis problem [1,16,17]. As far as we know, this viewpoint is different from the behavior scheduling concept used in conventional methods [7,21,27].

As described before, most gestures have a temporal structure with multiple consecutive movement phases including a preparation (P) phase, a stroke (S) phase, and a retraction (R) phase [27]. Only the S phase is essential, as it is the most energetic and meaningful phase of the gesture. In this paper, the parts before and after the S phase are denoted as the P and R phases, respectively. Please see Fig. 1 as a reference.

3.1   Investigation by Case Study

In this section, we look for the guidelines for our synchronization algorithm through a case study. Consider a scenario in which a virtual agent talks with you about your diet when you eat ice cream on two days, which may exceed your preferred calorie consumption. The reason why we choose this scenario is that such persuasion and guidance in the scenario is very important in education applications. We prepare the following two scripts, with similar contents but different lengths and emotional states. Therefore, it is reasonable to use the same gestures but need to modify the gesture for synchronization. Note that both scripts are translated from Japanese. Only Japanese versions are used in the case study.

1. (for Day One) Good morning. Ah–, you must have eaten ice cream last night! *You solemnly promised to go on a diet!* Well, you gotta do some walking today. Since the weather is fine, let's go now.
2. (for Day Two) Good morning. Ah—, you must have eaten ice-cream again last night. That's two days in a row!! *Were you serious when you promised to go on a diet?* Well, now you gotta walk that much farther. The forecast says rain this afternoon, so let's go now.

The speech is recorded by a narrator. A staff member creates the animations manually using the authoring tool MikuMikuDance (no relation to our authoring tool), in which the facial expression and body pose are independently edited in each key-frame after loading a suitable gesture from a motion capture database. First, the staff member manually creates the animation for Script #1 sentence by sentence and uses it directly in Script #2 for the same block of sentences. Then, the staff member manually improves the animation for Script #2. With this processing, we analyze the important issues in manual operation by noting the differences among the animations. Firstly, we observe that our staff member needs to modify the *duration* of gestures to fit with the speech. For example, the punching gesture is used for the italic parts in both scripts to express the emotion of the agent (as shown in Fig. 8). The cycle number of the punching gesture for Script #1 is four while it is changed to six for Script #2. Secondly, we observe that our staff member changes the *timing* of the gesture to fit with the speech. For example, the peak of the hand-lifting motion is arranged to match with the word "let's go". The creation procedure infers an engineering approach to automatic synchronization of gesture and speech, i.e., a guideline of our algorithm. Note that it takes a lot of time (more than two hours for about 30-second animation) for our staff member to synchronize the gestures with the speech although the creation time is dependent on the creator's experience.

3.2 Synchronization Algorithm

Discovering the algorithm-friendly guideline in Sect. 3.1 let us view the synchronization problem as a motion synthesis problem, with both duration and timing issues satisfied automatically as the constraints of our algorithm. Based on the fact that many gestures are cyclic or have similar poses, we adopt the motion graph technique [1, 16, 17] in this paper upon considering the gesture's structure. Furthermore, we extend the original definition of motion graph by adding the meta data like edge weights and edge lengths, which is specially designed for our task. By using the motion graphs, a new motion can be synthesized efficiently using dynamic programming algorithm [16, 45], which is applicable to our synchronization task. Then, we refine the animation in an additional step, resulting in a two-step synchronization algorithm.

Coarse Synchronization with Motion Graphs. Motion Graph Construction: Considering the structure of a gesture, we construct a graph structure in and only in the stroke (S) phase for each gesture in the database, as shown in Fig. 1. As many embodied agent systems do [21, 27], the labeling data for gesture structures are created manually, which is acceptable because the number of gestures is not excessive. Similar to [1, 16, 17], the motion graphs consist of nodes and edges. In addition, edge weights and edge lengths as defined in this paper are designed to measure the smoothness and duration of motion. All the key-frames $V = \{t^1, t^2, \cdots, t^N\}$ in the S phase are selected as nodes. Two neighboring key-frames t^i and t^{i+1} are connected by a uni-directional edge

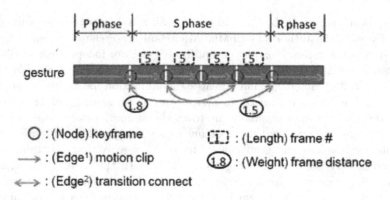

Fig. 1. A graph structure with node, edge, weight, and length is constructed in the stroke (S) phase of a gesture.

$e^{i,i+1}$, whose direction is the temporal direction. The edge weight $w^{i,i+1}$ of a uni-directional edge is zero and its edge length $L^{i,i+1}$ is the number of frames between the two nodes t^i and t^{i+1}. Two similar key-frames t^i and t^j that are not neighboring are connected by a bi-directional edge $e^{i,j}$, where $j \neq i \pm 1$ and the similarity or frame distance $d(t^i, t^j)$ is calculated as the weighted difference of joint orientations [43] as shown in Eq. (1).

$$d(t^i, t^j) = \sum_{m=1}^{M} w(m) \parallel \log(\mathbf{q}_{j,m}^{-1} \mathbf{q}_{i,m}) \parallel^2 \tag{1}$$

where M is the joint number, $w(m)$ denotes the joint weight, and $\mathbf{q}_{i,m}$ is the orientation of joint m in the i-th key-frame. The edge weight $w^{i,j}$ of a bi-directional edge is the above frame distance and its edge length $L^{i,j}$ is zero. For smooth transitions, motion blending is performed by the SLERP technique [39] for each bi-directional edge. Note that the above construction process can be performed off-line.

Search the Best Path: Given a sentence of script with timing tags and its speech, our system will show a list of gesture candidates that match the category of the text when the creator clicks it in the timeline. For example, "good morning" is a word in the greeting category, where gestures like bowing, hand waving, and light nodding are listed. This rule-based technique is popular for behavior selection in embodied agent systems [7,21]. Then the creator will select the best choice interactively. However, the original gesture motion in the database cannot always be a good match for the speech. In this section, our task is to generate a new motion for the given speech.

As [16] pointed out, any path in the motion graph is a new motion, and we search for the best one that best satisfies all the following conditions:

(1) as smooth as possible,
(2) the one with a length nearest to the desired duration L_{tg} that comes from the timing tags in the script,
(3) good connections with the P and R phases. Dynamic programming provides an efficient algorithm for motion graphs to search for the best path [45].

Basically, edge weight is used in the cost function for Condition (1), i.e. $cost(e^{i,j}) = w^{i,j}$ where $w^{i,j}$ denotes the edge weight from the i-th key-frame to the j-th key-frame, which may be a uni-directional edge or a bi-directional edge. For Condition (2), we check the cumulative length for the desired duration L_{tg} as Eq. (4) shows. For Condition (3), we set the initial node as the first key-frame t^1 in the S phase as Eq. (2) shows, which makes a natural connection to the P phase. In addition, we require the last node in the path is the last key-frame t^N in the S phase for good connection to the R phase. Finally, we select the best path with minimal cost that can satisfy all the three conditions. This makes the best path a new stroke phase with high quality and the desired duration.

$$P(t^v, 1) = \begin{cases} 0 & if \quad t^v = t^1 \\ \infty & others \end{cases} \tag{2}$$

$$P(t^v, k) = \min_{t^i \in V} \{ P(t^i, k-1) + cost(e^{i,v})) \} \tag{3}$$

$$P^* = \min_{L(P(t^N,k)) \geq L_{tg}} \{ P(t^N, k) \} \tag{4}$$

where $P(t^v, k)$ denotes the cost of the best path with k nodes and the last node of t^v, P^* denotes the best path for the speech, and $L(P(t^N, k))$ denotes the cumulative length of the best path $P(t^N, k)$.

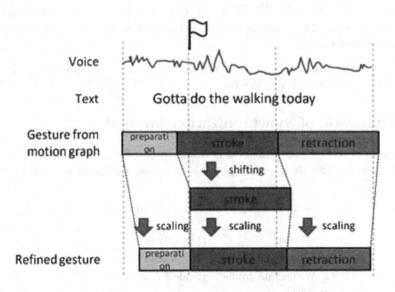

Fig. 2. Refine the synchronization by shifting and scaling the gesture motion, where the black flag shows the desired timing in the speech.

Fine Synchronization with Shifting and Scaling. In this section, we further improve the accuracy of synchronization with a shifting and scaling operation. As shown in Fig. 2, first the stroke (S) phase is shifted to the desired timing in the speech. Then, the S phase is scaled to match the desired duration as nearly as possible. In order to keep the motion natural and evocative of the desired emotional state, the scaling factor is limited to the range from 0.9 to 1.1. The same scaling factor will be used in the preparation (P) and retraction (R) phases to keep the motion consistency.

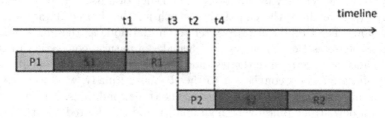

Fig. 3. Total scheduling of gestures in the timeline.

Total Scheduling. In most cases, our two-step algorithm will handle the duration and timing issues well. However, due to the preparation and retraction phases, neighboring gestures may conflict with each other as shown in Fig. 3. We define a rule to avoid such conflicts as shown in Table 1, where the conditions are:

(1) $t2 \leq t3$ (which means there is no conflict between R1 and P2),
(2) $t2 > t3$ & $\frac{(t2-t1)+(t4-t3)}{t4-t1} \leq TH$ (which means there is a conflict between R1 and P2 but the conflict is not serious),
    and (3) $t2 > t3$ & $\frac{(t2-t1)+(t4-t3)}{t4-t1} > TH$ (which means there is a conflict between R1 and P2 and the conflict is too serious to use the scaling operation), respectively.

Note that $t1$, $t2$, $t3$, and $t4$ are marked in Fig. 3. TH is a threshold, which is decided by preliminary experiments.

4   Evaluation of Synchronization Method

Evaluation of the proposed method is challenging. Objective evaluation of synchronization is difficult if not impossible because it is difficult to define the

Table 1. Total scheduling to avoid conflicts.

Conditions	Operations
(1) No conflict	No change
(2) With slight conflict	Scale P2 & R1 phases
(2) With serious conflict	Remove P2 & R1 phases

ground truth of synchronization. Currently, subjective evaluation is the only option as the conventional method [21] did.

Because our target user is the general consumer, the participants are non-expert volunteers. In our experiments, eleven participants are asked to evaluate two kinds of contents including the diet scenario (see details in Sect. 3.1) and a news comment scenario that comes from the comment session of a Japanese news site (see details in Table 2). The animation of the diet scenario lasts about 30 s, the news comment scenario about 5 s. The 11 participants, 5 male and 6 female, range in age from their 30 s to their 50 s, most with little experience or knowledge of animation creation.

Table 2. Scripts of Twitter comments on a news item about a company's new software, which are used in subjective evaluation.

Comment	Script (Translated from Japanese)
#1	No plan to release it, although they have a Japanese version!
#2	No plan to release a Japanese version! (The company) cannot seem to get motivated

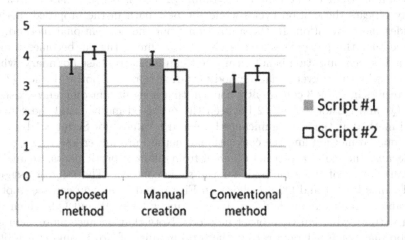

Fig. 4. Mean opinion scores for Q1 in diet scenario. The standard errors are shown in black lines (±1 SE).

For the diet scenario, three animations are shown that come (in random order) from manual creation, the proposed method, and the conventional method where the only difference from the proposed method is that the synchronization algorithm comes from [21]. Two questions (Q1: How good is the animation quality? Q2: How good is the synchronization of gesture and speech?) are evaluated using the following rating scheme. 5: Excellent; 4: Good; 3: Fair; 2: Poor; 1: Bad. The mean opinion scores (MOS) for Q1 and Q2 are listed in Figs. 4 and 5 respectively, where our method performs better in all cases than the conventional method, and in most cases is better than manual creation. Especially in

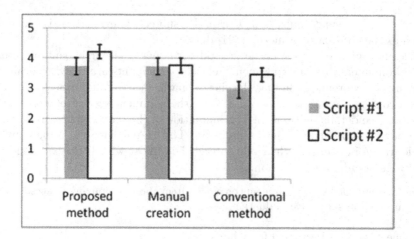

Fig. 5. Mean opinion scores for Q2 in diet scenario. The standard errors are shown in black lines (±1 SE).

Script #2, our method performs much better than manual creation, receiving a MOS of more than 4 (Good) in both Q1 and Q2. This is because manual creation simply repeats the motion cycles while the best path in the proposed method provides more variation. In the animation from the conventional method, we observe that the gesture's speed is changed so much that it becomes unnatural, and synchronization is not clear due to an unsharp phase boundary, which explains why the conventional method performs worst for both Q1 and Q2. T-test[2] results in Table 3 consistently confirm that a significant difference exists in both Q1 and Q2 for Script #2 between the proposed method and the conventional method at the 5 % significance level and p values[3] for Script #1 between the proposed method and the conventional method are rather low.

Because the scores are rather different from different participants, we analyze the rank rating of three methods for the same contents. The tally of assigned No. 1 ranks for Q1 and Q2 are shown in Figs. 6 and 7. As you can see, 9 of 11 participants rank our method No. 1 for Script #2 for Q1 and Q2, which is a much better evaluation than other methods. Note that because more than one method may get No. 1 rank rating, the total number of No. 1 ranking is a little more than 11 as shown in Figs. 6 and 7.

For the news comment scenario, we ask the participants to choose the better animation: the one produced by our proposed method or one by the conventional method (presented in random order) in terms of the two criteria above. For Comments #1 and #2, 8 of 11 participants select the animation produced using our proposed method. Five participants select our proposed method in both comments, while none selects the conventional method in both comments. This suggests that the proposed method is very promising.

---

[2] T-test is most commonly applied to determine if two sets of data are significantly different from each other when the samples follow a normal distribution.

[3] A lower p value means a higher probability that a significant difference exists.

Table 3. p values in T-test for diet scenario. PM: Proposed method vs. Manual creation. PC: Proposed method vs. Conventional method. Red fonts: $p < 0.05$. Blue fonts: $p < 0.1$.

		PM	PC
Script #1	Q1	0.4131	0.1443
	Q2	0.9817	0.0690
Script #2	Q1	0.1469	0.0422
	Q2	0.2050	0.0356

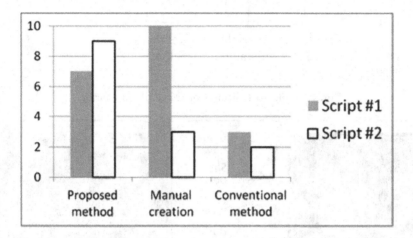

Fig. 6. Rank No.1 ratings for Q1 in diet scenario.

5 Overview of Authoring Tool

In this section, we will briefly introduce our system, an authoring tool whose interface is shown in Fig. 8. The purpose of the authoring tool is to provide a good balance between the creator's flexibility and his/her efficiency with a user-friendly user interface. For example, with our authoring tool, a staff member can create attractive animation for lectures even if he/she has little knowledge of animation creation.

User-friendly UI: There are five zones in our authoring tool. The top left window (Zone #1) in Fig. 8 is an affective space panel based on the circumplex model [37], where the creator can select a point and see the resulting facial expression in the top right window (preview window, Zone #3) in real time. The middle left window (Zone #0) is the control panel, where the creator can input the audio file and switch the display modes for the preview window and the candidate list window (Zone #2, top middle window in Fig. 8). The bottom window (Zone #4) shows the timeline for text, affect, facial expression, and gesture, respectively, which can be saved as a project file or loaded from a project file.

Efficiency: The creation procedure of a new project is as follows, which consists of gesture operations and facial operations. Before starting operation, an

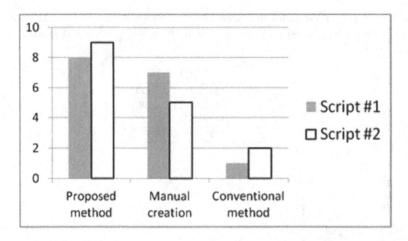

Fig. 7. Rank No.1 ratings for Q2 in diet scenario.

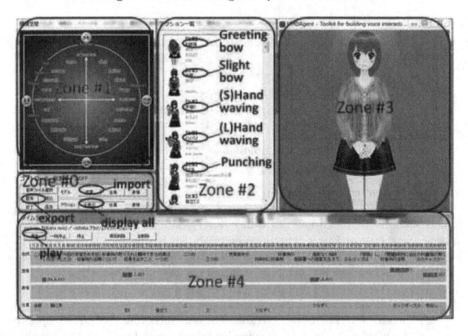

Fig. 8. User interface of our authoring tool.

audio file for narration or speech synthesis and its text file with timing tags are loaded. The text will be displayed in the time-line (Zone #4). There are only two clicks in each gesture operation. Firstly, when a user clicks a specific speech segments in the timeline, its audio is played, and at the same time some candidate gestures are listed (Zone #2) based on word matching between the speech texts and symbolized tags on each gesture (these tags are stored in a database with all motions). Secondly, when the user clicks the best gesture in the list (Zone #2) to fit the speech contents, our tool automatically modifies the original motions in a preset database in real time for synchronization with the speech, which uses the algorithm in Sect. 3. If the user does not satisfy the listed gestures, the user can select a new one from all gestures in the database by pressing the "display all" button. On the other hand, in a facial operation, the user clicks a point on the circumplex model of affect (Zone #1) to fit the emotion within the speech. A facial expression with that emotion is then generated based on the Facial Action Coding System (FACS) technique [9], where a total of 19 action units are defined in the case of our tool. After repeating two operations for all desired segments, the user can watch the entire animation by pressing the "play" button.

Flexibility: To re-use a large amount of CGM (Consumer Generated Media) resources in characters and motions on the Internet, we use the MikuMiku-Dance format (http://www.geocities.jp/higuchuu4/index_e.htm) in our system, which is very successful in Japan and East Asia[4]. Our authoring tool can import the character model data with MikuMikuDance format by simply pressing the "import" button and export the entire animation data with that format by pressing the "export" button. By then inputting the animation data into Miku-MikuDance software used in Sect. 3.1, the user can create a movie of an embodied agent with attractive animation. At the same time, the user can also add lecture slides such as Microsoft PowerPoint to the animation. In addition, using some free software, the data in MikuMikuDance format can be transferred from/to other formats like Maya and Blender.

As a result, it takes much less time to create animations with our authoring tool than by manual creation. Although the creation time is dependent on the experience of creator (which infers that the detailed time data have little meaning), it only takes about 10 % of time for our staff member to create the contents used in Sect. 4 by our authoring tool. Moreover, in our opinion, their quality is almost the same.

6   Applications to Education and Effectiveness Evaluation

Using our authoring tool, our staff creates several animations (i.e. 12 short animations in Experiment #1 and 2 long animations in Experiment #2) for digital education (See examples in Figs. 9 and 10). In this section, two experiments are conducted as follows. The purpose of Experiment #1 is to find a best practice for embodied agents, especially the effectiveness of facial expressions and gestures. Using the best practice, in Experiment #2, we then compare the effectiveness of embodied agent with video and no agent.

Experiment #1: To avoid the possible bias from different levels of students' knowledge, short animations in four different categories are created and

---

[4] In this paper, the so called MikuMikuDance has two meanings. First, it may mean the authoring tool used in Sect. 3.1. Second, it may mean the specifications for mesh model, motion, and other data in animation.

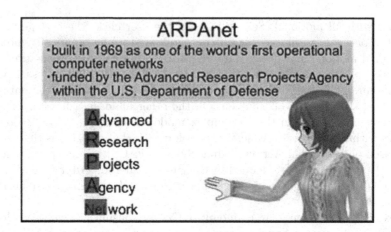

Fig. 9. Screen capture of an education animation in Experiment #1. Note that the text in the slide is translated from Japanese. Only Japanese versions are used in the evaluation.

evaluated by each student including information technology, history, chemistry, and geography. In each content, there is a short lecture slide with an embodied agent that lasts about one minute and a test with three questions including a question about figures. By selecting different facial expressions and gestures, three different styles of animations are evaluated, which include no expression, moderate expression, and intensive expression or no gesture, regular gestures, and highlighted gestures. As Table 4 shows, more gestures and stronger gestures are used in the style of highlighted gestures to emphasize the key points in the education contents.

Table 4. Statistics for different gesture styles.

Gesture style	No gesture	Regular gestures	Highlighted gestures
Average number of gestures	0.0	9.7	15.3
Gesture example	-	point a finger up	beat on the slide

Basically, we want to see the effectiveness of agent by the scores (percentage of correct answers) in the test after watching the animations. Totally 34 participants conduct the evaluation on the animations in four categories, whose styles of gesture and expression are randomly selected. The average scores for facial expressions and gestures are shown in Tables 5 and 6 respectively, which were averaged by all the participants in all four categories. The results show that both the highlighted gestures and intensive expressions are effective to obtain better scores. Especially, a high score of 72.7 % is obtained in the questions about figures for the highlighted gestures. A T-test is performed between no gesture and highlighted gestures, which gives a significant level of 6.96 %. A similar T-test between no expression and intensive expressions gives a significant level of

Table 5. Average scores (percentage of correct answers) for different gesture styles.

Gesture style	No gesture	Regular gestures	Highlighted gestures
Average scores for all questions	56.7%	60.7%	66.7%
Average scores for questions about figures	57.4%	60.0%	72.7%

Table 6. Average scores (percentage of correct answers) for different expression styles.

Expression style	No expression	Moderate expressions	Intensive expressions
Average scores for all questions	56.3%	60.0%	67.4%
Average scores for questions about figures	55.6%	66.7%	67.4%

7.74%. Both are rather near to 5%, which is commonly used as a significant difference. In addition, many participants report that the intensive expressions and highlighted gestures are impressive in the questionnaire.

Experiment #2: In a real situation of the "Web Management" course, we record the professor's speech and create the digital material from a university lecture with the agent and its synchronized attractive animations using our authoring tool as shown in Fig. 10. At the same time, we also record a video lecture with a real human as shown in Fig. 11 and the contents without the agent or real human for comparison. Each type of contents consists of the first part and second part where each part lasts about six minutes. Three different types (i.e. video lectures, no agent, and agent animation) of contents are evaluated by each student, resulting in six pieces of test contents. Moreover, to avoid possible bias due to different levels of student knowledge, they also watch different contents between the first and second parts (This is the reason why we divide the lecture into two parts).

An evaluation of the contents was conducted with a total of 32 participants from the same university and in which content types were randomly selected. The average scores for the test are shown in Table 7, which were averaged by all the participants and all the questions. The results show that both animation contents with an agent and without an agent are as effective in obtaining better scores as a video lecture. Because the test scores were a part of the final scores of the course, most of the students took notes carefully, which led to no difference among three methods. On the other hand, the scores in the second part were considerably lower than them in the first part. We believe this is because the student's concentration decreased in the latter part. The results of questionnaire are shown in Fig. 12 and Table 8. In Fig. 12, the question of "Do you

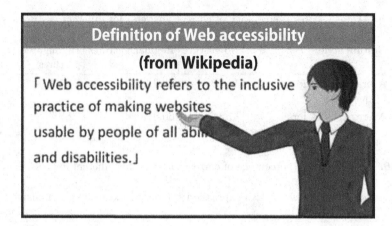

Fig. 10. Screen capture of an education animation in Experiment #2. Note that the text in the slide is translated from Japanese. Only Japanese versions are used in the evaluation.

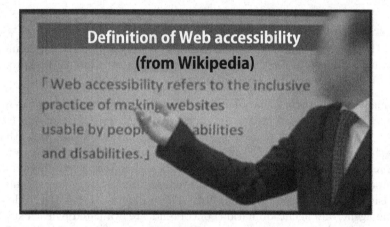

Fig. 11. Screen capture of a recorded video in Experiment #2. Note that the text in the slide is translated from Japanese. Only Japanese versions are used in the evaluation. Professor's face is blurred for publication but the students watched a clear video.

want to use these contents?" is evaluated using the following rating scheme. 4: Definitely Yes, 3: Rather Yes, 2: Rather No, 1: Definitely No. Because over 70 % of the participants selected 3 or 4, the agent animation is definitely more effective than other types for improving student motivation. In Table 7, in response to the question of "Did you enjoy studying it?", the percentage of participants answering "Yes" or "No" is shown. Since the percentage of "No" was less than 40 % in the second part, the agent animation is therefore effective in sustaining the student's motivation.

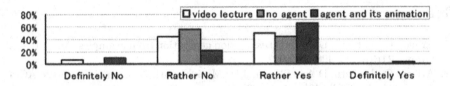

Fig. 12. Results of questionnaire for the question "Do you want to use these contents?".

Table 7. Average scores for different content types.

Content type	Video lecture	No agent	Agent animation
First part	71.3 %	68.8 %	69.4 %
Second part	27.5 %	47.5 %	46.3 %
Whole	49.4 %	58.1 %	57.8 %

Table 8. Percentage of participants answering "Yes" or "No" for "Q: Did you enjoy studying?".

Content type	Video lecture		No agent		Agent animation	
Answer	Yes	No	Yes	No	Yes	No
First part	75.0 %	25.0 %	100 %	0	93.7 %	6.3 %
Second part	37.5 %	62.5 %	25.0 %	75.0 %	62.5 %	37.5 %

7   Conclusions and Future Work

In this paper, we have described an authoring tool we have implemented to facilitate the creation of high quality animation for embodied agents, with facial expressions and gestures that are accurately synchronized with speech. In a precedent case study, we have investigated the important issues (i.e. duration and timing) in synchronizing of gesture with speech, which has led us to consider the synchronization problem to be a motion synthesis problem. We have proposed a novel two-step solution using the motion graph technique within the constraints of gesture structure. Subjective evaluation of two scenarios involving talking and news commentary has demonstrated that our method is more effective than the conventional method. Furthermore, we have conducted several experiments in a university's class, where we get rather positive feedback from users.

In the future, we plan to improve the generation of facial expressions, where realistic facial dynamics can further improve animation quality. At the same time, we are extending the target applications to new categories such as remote chat and human-robot interaction.

Acknowledgments. All the participants in our experiments, especially Prof. Shirotomo Aizawa and his students in Nagoya University of Arts and Sciences, Japan, are greatly appreciated.

References

1. Arikan, O., Forsyth, D.: Interactive motion generation from examples. ACM Trans. Graph. **21**(3), 483–490 (2002)
2. Arikan, O., Forsyth, D.A., O'Brien, J.F.: Motion synthesis from annotations. ACM Trans. Graph. **22**(3), 402–408 (2003)
3. Beaudoin, P., Coros, S., van de Panne, M., Poulin, P.: Motion-motif graphs. In: SCA 2008, pp. 117–126 (2008)
4. Beskow, J., Engwall, O., Granstrom, B., Wik, P.: Design strategies for a virtual language tutor. In: INTERSPEECH-2004, pp. 1693–1696 (2004)
5. Cassell, J., Bickmore, T., Billinghurst, M., Campbell, L., Chang, K., Vilhjálmsson, H., Yan, H.: Embodiment in conversational interfaces: Rea. In: CHI 1999, pp. 520–527 (1999)
6. Cassell, J., Sullivan, J., Prevost, S., Churchill, E.F.: Embodied Conversational Agents, 1st edn. The MIT Press, Cambridge (2000)
7. Cassell, J., Vilhjálmsson, H.H., Bickmore, T.: Beat: the behavior expression animation toolkit. In: ACM SIGGRAPH 2001, pp. 477–486 (2001)
8. Dutoit, T.: An Introduction to Text-to-Speech Synthesis. Springer, New York (2001)
9. Ekman, P., Friesen, W.V., Hager, J.C.: Facial Action Coding System: The Manual on CD ROM. A Human Face, Salt Lake City (2002)
10. Forsyth, D.A., Arikan, O., Ikemoto, L., O'Brien, J.F.: Computational studies of human motion: Part 1, tracking and motion synthesis. Found. Trends Comput. Graph. Vis. **1**(2), 77–254 (2006)
11. Gleicher, M., Shin, H.J., Kovar, L., Jepsen, A.: Snap-together motion: assembling run-time animations. In: I3D 2003, pp. 181–188 (2003)
12. Gulz, A., Haake, M., Silvervarg, A.: Extending a teachable agent with a social conversation module – effects on student experiences and learning. In: Biswas, G., Bull, S., Kay, J., Mitrovic, A. (eds.) AIED 2011. LNCS, vol. 6738, pp. 106–114. Springer, Heidelberg (2011)
13. Huang, J., Pelachaud, C.: Expressive body animation pipeline for virtual agent. In: Nakano, Y., Neff, M., Paiva, A., Walker, M. (eds.) IVA 2012. LNCS, vol. 7502, pp. 355–362. Springer, Heidelberg (2012)
14. Ieronutti, L., Chittaro, L.: Employing virtual humans for education and training in X3D/VRML worlds. Comput. Educ. **49**(1), 93–109 (2007)
15. Kopp, S., Krenn, B., Marsella, S.C., Marshall, A.N., Pelachaud, C., Pirker, H., Thórisson, K.R., Vilhjálmsson, H.H.: Towards a common framework for multimodal generation: the behavior markup language. In: Gratch, J., Young, M., Aylett, R.S., Ballin, D., Olivier, P. (eds.) IVA 2006. LNCS (LNAI), vol. 4133, pp. 205–217. Springer, Heidelberg (2006)
16. Kovar, L., Gleicher, M., Pighin, F.: Motion graphs. ACM Trans. Graph. **21**(3), 473–482 (2002)
17. Lee, J., Chai, J., Reitsma, P., Hodgins, J., Pollard, N.: Interactive control of avatars animated with human motion data. ACM Trans. Graph. **21**(3), 491–500 (2002)
18. Lee, J., Lee, K.H.: Precomputing avatar behavior from human motion data. In: Proceedings of the 2004 ACM SIGGRAPH/Eurographics symposium on Computer animation, pp. 79–87 (2004)
19. van Luin, J., op den Akker, R., Nijholt, A.: A dialogue agent for navigation support in virtual reality. In: CHI EA 2001, pp. 117–118 (2001)

20. Maldonado, H., Lee, J.E.R., Brave, S., Nass, C., Nakajima, H., Yamada, R., Iwamura, K., Morishima, Y.: We learn better together: enhancing elearning with emotional characters. In: CSCL 2005, pp. 408–417 (2005)
21. Marsella, S., Xu, Y., Lhommet, M., Feng, A., Scherer, S., Shapiro, A.: Virtual character performance from speech. In: SCA 2013, pp. 25–35 (2013)
22. McGurk, H., MacDonald, J.: Hearing lips and seeing voices. Nature **264**, 746–748 (1976)
23. McNeill, D.: Gesture and Thought. University of Chicago Press, Chicago (2005)
24. McNeill, D.: So you think gestures are nonverbal? Psychol. Rev. **92**(3), 350–371 (1985)
25. Miller, L.M., D'Esposito, M.: Perceptual fusion and stimulus coincidence in the cross-modal integration of speech. J. Neurosci. **25**(25), 5884–5893 (2005)
26. Mizuguchi, M., Buchanan, J., Calvert, T.: Data driven motion transitions for interactive games. In: Proceedings of EUROGRAPHICS 2001 short papers (2001)
27. Neff, M., Kipp, M., Albrecht, I., Seidel, H.P.: Gesture modeling and animation based on a probabilistic re-creation of speaker style. ACM Trans. Graph. **27**(1), 5:1–5:24 (2008)
28. Ng-Thow-Hing, V., Luo, P., Okita, S.: Synchronized gesture and speech production for humanoid robots. In: IEEE/RSJ IROS 2010, pp. 4617–4624 (2010)
29. Niewiadomski, R., Bevacqua, E., Mancini, M., Pelachaud, C.: Greta: an interactive expressive ECA system. In: AAMAS 2009, vol. 2. pp. 1399–1400 (2009)
30. Nishida, T.: Conversational Informatics: An Engineering Approach. Wiley, New York (2007)
31. Noma, T., Zhao, L., Badler, N.: Design of a virtual human presenter. IEEE Comput. Graph. Appl. **20**(4), 79–85 (2000)
32. Ogan, A., Finkelstein, S., Mayfield, E., D'Adamo, C., Matsuda, N., Cassell, J.: "oh dear stacy!": Social interaction, elaboration, and learning with teachable agents. In: CHI 2012, pp. 39–48 (2012)
33. Oura, K., Yamamoto, D., Takumi, I., Lee, A., Tokuda, K.: On-campus, user-participatable, and voice-interactive digital signage. J. Jpn Soc. Artif. Intell. **28**(1), 60–67 (2013)
34. Reitsma, P.S.A., Pollard, N.S.: Evaluating motion graphs for character animation. ACM Trans. Graph. **26**(4), 18 (2007)
35. Ren, C., Zhao, L., Safonova, A.: Human motion synthesis with optimization-based graphs. Comput. Graph. Forum **29**(2), 545–554 (2010)
36. Rist, T., Andr, E., Baldes, S., Gebhard, P., Klesen, M., Kipp, M., Rist, P., Schmitt, M.: A review of the development of embodied presentation agents and their application fields. In: Prendinger, H., Ishizuka, M. (eds.) Life-Like Characters. Cognitive Technologies, pp. 377–404. Springer, Berlin (2004)
37. Russell, J.A.: A circumplex model of affect. J. Pers. Soc. Psychol. **39**(6), 1161–1178 (1980)
38. Safonova, A., Hodgins, J.K.: Construction and optimal search of interpolated motion graphs. ACM Trans. Graph. **26**(3), 106 (2007)
39. Shoemake, K.: Animating rotation with quaternion curves. In: ACM SIGGRAPH 1985, pp. 245–254 (1985)
40. Soliman, M., Guetl, C.: Intelligent pedagogical agents in immersive virtual learning environments: a review. In: MIPRO 2010, pp. 827–832 (2010)
41. Stone, M., DeCarlo, D., Oh, I., Rodriguez, C., Stere, A., Lees, A., Bregler, C.: Speaking with hands: creating animated conversational characters from recordings of human performance. ACM Trans. Graph. **23**(3), 506–513 (2004)

42. Čerekovič, A., Pandžič, I.: Multimodal behavior realization for embodied conversational agents. Multimedia Tools Appl. **54**(1), 143–164 (2011)
43. Wang, J., Bodenheimer, B.: An evaluation of a cost metric for selecting transitions between motion segments. In: SCA 2003, pp. 232–238 (2003)
44. Xu, J., Myodo, E., Sakazawa, S.: Motion synthesis for affective agents using piecewise principal component regression. In: IEEE ICME 2013, pp. 1–7 (2013)
45. Xu, J., Takagi, K., Sakazawa, S.: Motion synthesis for synchronizing with streaming music by segment-based search on metadata motion graphs. IEEE ICME 2011, pp. 1–6 (2011)
46. Zhao, L., Safonova, A.: Achieving good connectivity in motion graphs. Graph. Models **71**(4), 139–152 (2009)

Abstraction of Heterogeneous Supplier Models in Hierarchical Resource Allocation

Alexander Schiendorfer[1]([✉]), Gerrit Anders[1], Jan-Philipp Steghöfer[2], and Wolfgang Reif[1]

[1] Institute for Software and Systems Engineering,
University of Augsburg, Augsburg, Germany
{schiendorfer,anders,reif}@isse.de
[2] University of Gothenburg, Gothenburg, Sweden
jan-philipp.steghofer@cse.gu.se

Abstract. Resource allocation problems such as finding a production schedule given a set of suppliers' capabilities are generally hard to solve due to their combinatorial nature, in particular beyond a certain problem size. Large-scale instances among them, however, are prominent in several applications relevant to smart grids including unit commitment and demand response. Decomposition constitutes a classical tool to deal with this increasing complexity. We present a hierarchical "regio-central" decomposition based on abstraction that is designed to change its structure at runtime. It requires two techniques: (1) synthesizing several models of suppliers into one optimization problem and (2) abstracting the direct composition of several suppliers to reduce the complexity of high-level optimization problems. The problems we consider involve limited maximal and, in particular, minimal capacities along with on/off constraints. We suggest a formalization termed *supply automata* to capture suppliers and present algorithms for synthesis and abstraction. Our evaluation reveals that the obtained solutions are comparable to central solutions in terms of cost efficiency (within 1 % of the optimum) but scale significantly better (between a third and a half of the runtime) in the case study of scheduling virtual power plants.

Keywords: Hierarchical resource allocation · Self-organization · Discrete optimization · Abstraction · Virtual power plants

1 Dynamic Resource or Task Allocation in a Hierarchical Setting

Resource or task allocation problems present themselves in a variety of domains addressed by multi-agent-systems [7], including distributed power management [11,36] or grid computing [1]. In this paper, we investigate resource allocation

---

This research is partly sponsored by the German Research Foundation (DFG) in the project "OC-Trust" (FOR 1085).

N.T. Nguyen et al. (Eds.): TCCI XX, LNCS 9420, pp. 23–53, 2015.
DOI: 10.1007/978-3-319-27543-7_2

problems with a demand imposed by the environment that needs to be satisfied by a set of suppliers which are represented by software agents acting on their behalf. These problems also lie at the heart of the selection and payment function of several practical problems addressed by mechanism design including adaptations of the Vickrey-Clarke-Groves mechanism [11,44] where computationally efficient algorithms are required. These problems can, in general, be expressed as constraint satisfaction and optimization problems. Techniques from operations research and discrete optimization, such as constraint programming or mixed integer (linear) programming, have been proposed to find optimal solutions (see, e.g., [19,37]). This proves particularly useful when the characteristics of heterogeneous suppliers requiring different sets of variables and constraints need to be modeled.

With regard to power management systems, a specific resource allocation problem is to maintain the balance between production and consumption at all times to keep the mains power frequency in a small corridor to achieve stable power supply [18]. This is achieved by creating "schedules" for controllable power plants, i.e., instructions of how much power they have to produce at which point in time, based on the predicted demand and the predicted input of weather-dependent power plants at that time. A complicating factor is the suppliers' inability to switch production levels arbitrarily over time. In this context, three main challenges arise: First, the resource allocation problem involves a vast number of power plants and consumers but, at the same time, has to be solved in a timely fashion. Second, the balance has to be kept despite uncertain demand and output of weather-dependent power plants. Third, heterogeneity requires solutions that can deal with the power plants' individual characteristics in the form of technical limitations and preferences when creating schedules.

If the size of the system prohibits a centralized solution, either due to the communication overhead required in collecting all necessary information or due to the complexity of a centralized solution model, hierarchical decomposition offers a generic tool to deal with these issues (see, e.g., [1,6,14]). Here, the global problem is decomposed by forming a hierarchical structure of agent organizations [20,43]. To solve the global problem, each organization acts as an *intermediary* that has to recursively solve a sub-problem, which is achieved in a top-down fashion with regard to the hierarchy, as illustrated in Fig. 1. Given a hierarchy, the problem can be solved by means of an auction protocol [2] relying only on proposals submitted by subordinates. Alternatively, intermediaries can centralize information from a region of the system, i.e., the control models of their subordinates, and solve their sub-problem centrally. Hence, we term this approach *regio-central*. Independent sub-problems can be solved concurrently in both endeavors. Lacking global knowledge, overall proven optimality is generally not feasible but traded for tractable sub-problems and close-to-optimal solutions (as evidenced by our evaluation results in Sect. 5.1). To achieve this tractability, we propose algorithms to calculate *abstractions* [15]:

Abstraction: On higher levels, intermediaries decide on schedules that have to be fulfilled by their subordinate agents. To make optimal decisions, exact models

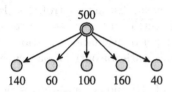
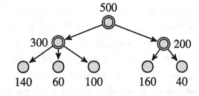

Fig. 1. A central and hierarchical solution example to a resource allocation problem. Inner nodes representing intermediaries are marked by double circles and redistribute their assigned share of an overall demand, e.g., power.

of all subordinates in composition would be required—eventually yielding the same complexity as a centralized solution. Because of the dynamic composition of organizations, abstraction has to be performed at runtime without manual interference.

In this paper, we focus on the regio-central approach to evaluate abstraction techniques in isolation, effectively testing the quality of the calculated models. However, abstracted models are also employed in the auction-based algorithm [3], making them a prerequisite for our hierarchical approach.

In the light of a dynamic environment and uncertainties, an organization might turn out to be unable to solve its corresponding sub-problem with sufficient quality or in due time, though. The organizations might thus have to or prefer to adapt their composition at runtime [3]. Furthermore, it is far from obvious how suitable structures have to be defined in terms of the size of organizations and the depth of the hierarchy in order to yield minimal runtimes (a challenge we discuss in our evaluation in Sect. 5.3) and solutions of acceptable quality at the same time. The size of the search space of possible hierarchies is $\Theta(B_n)$ where B_n is the nth Bell number denoting the number of partitions since for each partition we can construct at least one corresponding hierarchy [4]. As a consequence, the system might thus depend on a number of successive reorganizations to establish a structure that provides an adequate runtime. In terms of optimization problems and models, this rises the challenge of separating concerns adequately, i.e., separating individual supplier characteristics from intermediaries' objectives, such that we can *synthesize* optimization problems at runtime:

Synthesis: To be able to solve the resource allocation problem at the different levels of the hierarchy considering the dynamic composition of organizations, *synthesis* defines necessary interfaces to convert a set of heterogeneous control models along with a predefined objective to obtain a regio-central optimization problem. The intermediary distributes the solution of this problem to its subordinate agents which, in turn, redistribute the solution if they are intermediaries themselves.

Throughout the paper, the creation of power plant schedules in autonomous power management systems serves as a case study to detail the life cycle and algorithms for this regio-central solution approach. We present a formalization of the resource allocation problem as a mixed integer program in Sect. 2 followed by considerations on how control models of suppliers are combined to one such problem in Sect. 3. Three algorithms to obtain abstract control models of intermediaries are motivated and presented in Sect. 4. In Sect. 5 we discuss our experimental setup and present results to a number of evaluation questions. We present the related work in Sect. 6 and conclude with an outlook on achieved results and open questions in Sect. 7.

The paper is a substantially revised version of [41] and emphasizes the problem of cost-efficient scheduling rather than minimizing deviations between supply and demand only. We added the formal definition of supply automata to enable the systematic generation of synthesized optimization problems for a variety of generators. Also, the approach has been re-evaluated on larger experiments and the results are presented more rigorously. Moreover, the algorithms are discussed in more detail accompanied by additional examples and complexity considerations.

2   Problem Formalization

In the context of power management as a representative for the considered problem, the resource to be allocated is electric power and an allocation specifies how much (controllable) power plants need to supply at a certain point in time. The task is to assign schedules to controllable power plants so that their joint output meets the *residual load*, i.e., the difference between demand and supply from intermittent, weather-dependent sources. The overall problem is to continuously keep this balance at all times.

Formally, we abstract from this continuity by considering a finite set of discrete time steps $\mathcal{T} = \{1, \ldots, t_{\max}\}$ (e.g., in intervals of 15 min as used by energy markets) where we measure discrepancies and collect predictions of expected demand and intermittent production. Nonetheless, the problem cannot be solved optimally in advance for, e.g., the next day, due to the dynamic and stochastic nature mentioned in Sect. 1. An optimal schedule under the assumption of a sunny day might drastically underestimate the residual load and thus be detrimental to the system's objective. However, we need to consider at least a certain scheduling window $\mathcal{W} \subset \mathcal{T}$ ranging from the next time step $t_{\text{now}} + 1$ up to $t_{\text{now}} + W$ with $W = |\mathcal{W}|$ being the fixed width of that window. Inertia manifested in physical constraints such as limited rates of change or start-up times—both factors that present in slow-changing thermal power plants—necessitates this proactive behavior [18]. For instance, if a power plant's production is necessary to meet a high demand at time $t_{\text{now}} + W$, it may be required to start ramping up in $t_{\text{now}} + 1$.

In the energy management setting, suppliers are further subject to both limited maximal and *minimal* capacities of production due to mostly technical but also economical reasons (e.g., a minimum generation of 20 % of the nameplate capacity for gas turbines or 40 % for coal based thermal plants [21]). Moreover, for a virtual power plant, it is imperative to be able to switch plants on and off selectively to achieve more favorable aggregate partial load behavior compared to a single conventional generator [24] that suffers from increased costs when not operated at optimal energy conversion efficiency. Based on these assumptions, we present the core optimization problem we consider in Eq. (1).

$$\underset{S_t^a, \sigma_t^a}{\text{minimize}} \qquad \alpha_\Delta \cdot \Delta + \alpha_\Gamma \cdot \Gamma \qquad\qquad\qquad (1)$$

subject to $\quad \forall a \in \text{sub}(v), \forall t \in \mathcal{W}:$

$$\exists [x, y] \in L^a : x \le S_t^a \le y,$$

$$S_\delta^{a-}\left(\sigma_{t-1}^a, S_{t-1}^a\right) \le S_t^a \le S_\delta^{a+}\left(\sigma_{t-1}^a, S_{t-1}^a\right)$$

with $\Delta = \sum_{t \in \mathcal{W}} \left| \sum_{a \in \text{sub}(i)} S_t^a - D_t \right|,$

and $\Gamma = \sum_{t \in \mathcal{W}, a \in \text{sub}(i)} \kappa_a(S_t^a)$

The problem is formulated for one intermediary i that controls its set of subordinate agents $\text{sub}(i)$. The task primarily consists of assigning scheduled contributions.

S_t^a for each agent $a \in \text{sub}(i)$ and time step $t \in \mathcal{W}$. Additional *state* of agents, such as it being on or off, is taken into consideration with the auxiliary variables σ_t^a. Two possibly conflicting objectives to rank schedules are given by minimizing the discrepancy Δ between the aggregate supply and the demand D_t given by the residual load as well as minimizing the cost for the scheduled supply Γ which depends on cost functions κ_a mapping production levels to costs. The prioritization of both goals is regulated by weights α_Δ and α_Γ.

Minimal and maximal production capacities $S_{\min}^a$ and $S_{\max}^a$ are generalized to finite lists of non-overlapping *feasible regions* L^a, representing minimal and maximal capacities in a particular *mode*, e.g., on or off. A single generator that can be switched off[1] is then represented by $L^a = \langle [0,0], [S_{\min}^a, S_{\max}^a] \rangle$. Section 4 sheds light on why the generalization to lists is necessary to deal with abstracted models of intermediaries representing a whole set of suppliers. Intuitively, the composition of n suppliers having 2 modes each leads to 2^n combined modes showing individual minimal and maximal boundaries.

Inertia is reflected by functions S_δ^{a-} and S_δ^{a+} restricting possible decreasing or increasing supply given a state and current supply. Their role is explored more thoroughly and exemplified in Sects. 3 and 4.2. A state σ_t^a can be seen as a (heterogeneous) store of variables depending on the type of a. For instance, a supplier showing minimal runtimes before switching off could store the current runtime in time steps as $\sigma_t^a.cr$. With respect to a scheduling window $\mathcal{W} = \{t_{\text{now}}+1, \ldots, t_{\text{now}}+W\}$, $\sigma_{t_{\text{now}}}^a$ contains the actual momentary state as constants, whereas for $t \in \mathcal{W}$, σ_t^a are decision variables that have to be assigned consistently by the optimizer (e.g., when to issue a start/stop signal). If a supplier $a \in \text{sub}(i)$ is an intermediary itself, it solves the identical optimization problem for its own subordinates $\text{sub}(a)$ with its assigned share S_t^a being the demand D_t to be redistributed. Guaranteeing that S_t^a is in fact achievable by $\text{sub}(a)$ requires calculating the schedules based on the composition of all underlying agents, making the problem utterly complicated to solve. To achieve a reduction in complexity, abstraction is employed instead (see Sect. 4).

---

[1] In the application domain, we also need to consider so-called must-run plants [48], leading to $L^a = \langle [P_{\min}^a, P_{\max}^a] \rangle$.

When allowing these optimization problems to be formulated dynamically at runtime, we need to separate concerns of individual suppliers and the optimization problem as a whole. A few terms help us to put these aspects in context. More specifically, they aid to decompose the challenge presented by the resource allocation problem into several subproblems:

Individual Agent Models (IAM) are variants of extended finite state machines ("supply automata", see Sect. 3) that describe the properties of *one* supplier, in particular possible supply trajectories over time as well as different modes.

Abstracted Agent Models (AAM) are approximations of the composition of a set of underlying agents and represent the joint behavior.

Synthesized Optimization Problems (SOP) combine *several* agent models with a predefined "template" objective to generate the optimization formulation presented in Eq. (1) at runtime. More specifically, the resulting SOP is a generated artifact in a suitable language, e.g., MiniZinc [32] or, in our case, the optimization programming language (OPL) [10].

3   Synthesis of Regio-Central Optimization Problems

The process of converting a set of control models into one optimization problem according to the template presented in Eq. (1) is termed "synthesis". We motivate the main ideas and techniques and refer to [42] for a reference implementation.

A driving factor is to model suppliers with heterogeneous physical requirements such that reconfiguration of the system's hierarchical structure at runtime is enabled. The suppliers we consider are in fact dynamical systems with a controller manipulating an underlying physical process to, e.g., ramp up, ramp down, or shut off the supplier. We search for a sufficiently detailed discrete-time model that we can employ in the scheduling problem to dismiss technically infeasible schedules yet abstract enough (e.g., not containing particular control signals) to solve the synthesized optimization problem efficiently.

First, we look at existing constraints in the power domain (see [5,34]) and discuss how to build ones appropriate for the problem we are concerned with. The types of constraints considered in the literature are:

Table 1. Cold and hot start-up times for different power plant types according to [23,30]. A cold start occurs if a plant is down for more than 48 h, a hot start if it is down for less than 8 h.

Plant type	Cold start-up (h)	Hot start-up (h)
Black coal	4 − 5	2
Brown coal	6 − 8	2 − 4
Gas turbine	0.5	0.25
Photothermal	4 − 5	2

Minimal up/down times: A supplier has to run (or be switched off) for a minimal number of steps before switching the mode from "on" to "off" or vice versa [34].

Ramp up/down rates: A fixed amount of ramp-up/ramp-down capacity between two consecutive time steps is assumed [45]; in thermal power plants, this is due to required heating and cooling.

Cold/warm start-up times: A thermal plant generally needs a minimum number of time steps to ramp up from 0 to its minimal production [5]. Depending on the supplier's type, this start-up duration depends on the down-time as "cold starts" differ from "warm starts" (see Table 1 for sample values).

Not all supplier types are subject to the *same* constraints. Consider, e.g., a small gas power plant where the ramp up/down rates are high enough to regulate from minimal to maximal production in one time step. Similarly, with respect to Table 1, the difference between a hot or cold start-up might be negligible (depending on the duration of one time step) for gas turbines but not for photothermal plants. Modeling this type of behavior requires storing the running or standing times as auxiliary variables and define their development over time by suitable relations.

In [5], the problem is solved by encoding the start-up using linear constraints and 0/1 decision variables directly in a MIP. We propose to take a short detour and first model a supplier as a variant off *extended finite state machines* (EFSM) that can be seen as a discrete-time variant of a rectangular hybrid automaton [17] and then *generate* decision variables and constraints in a suitable language, e.g., OPL for our synthesized CSOP. Different *modes* of operation correspond to, e.g., switching a plant on or off and private variables capture individual physical characteristics. If, for instance, starting-up takes several time steps, then turning a supplier on requires taking a transition from the off-mode to a dedicated start-up-mode (where several time steps have to be passed) before eventually arriving in the productive mode. Expressing the control models this way also paves the way for better modeling and debugging support by suitable tools such as simulations, reachability analysis ("Can a supplier reach $S_{\max}$ if started newly?"), or automated testing ("Is a given trajectory accepted by a supplier?") and offers a canonical transformation.

3.1 Supply Automata

A supply automaton[2] is a particular extended finite state machine that specifies how a supplier can change its output S as well as other *local* state variables over one time step. We use it to describe the set of feasible trajectories of a supplier.

---

[2] Supply is not limited to positive production, since negative production, e.g., consumption or storage of a resource, can be vital to meet the demand, reflecting the concept of a *prosumer*.

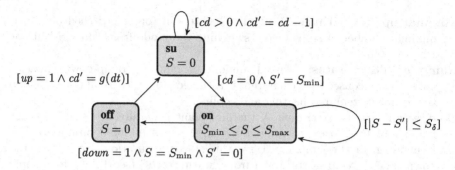

Fig. 2. Supply automaton to model adaptive start-up times depending on down time dt. A countdown cd is initialized with $g(dt) \in \mathbb{N}$, e.g., according to Table 1, and has the plant stay in the mode su (start-up) for $g(dt)$ steps. A plant can only contribute in the on-mode. Expressions in brackets contain jump predicates, invariants are written inside the modes.

Example 1. Figure 2 presents a supply automaton for a thermal unit considering start-up time. Consider a supplier that is currently off, needs constant $g(dt) = 1$ time steps to start up and then provide supply in $[5, 10]$ with a maximal change per time step $S_\delta = 1$. Then, exemplary trajectories $(0, 0, 5, 6, 7, 8, 9, 10)$ as well as $(0, 0, 0, 0)$ or $(0, 0, 5, 5, 0)$ are valid, in contrast to, e.g., $(0, 5, 5)$ which violates minimal start-up or $(0, 0, 5, 8)$ which violates maximal ramp-up.

We first explain the syntactic concepts used in Fig. 2 and provide semantics by defining valid transitions. We deliberately omit time indices in the presentation of supply automata as only transitions from one state to another are considered. The translation into an optimization problem similar to Eq. (1) including the underlying time series is presented in Sect. 3.2.

Definition 1. *A supply automaton SA is described by* $(M, X, T, \mathsf{Inv}, \mathsf{Jump})$ *where:*

- *M is a finite set of modes,*
- *X is a finite set of real-valued, local variables with $S \notin X$; we write $X_S = \{S\} \cup X$ to include S, the real-valued supply required for all suppliers. X'_S represents the same variables after a jump. $\Phi(X_S)$ denotes the set of predicates over free variables taken from X_S.*[3]
- *$T \subseteq M \times M$ indicate possible mode transitions*
- *$\mathsf{Inv} : M \to \Phi(X_S)$ returns the invariant in one mode including a feasible interval per mode "$x \leq S \leq y$" ($x \leq y \in \mathbb{R}$*
- *$\mathsf{Jump} : T \to \Phi(X_S \cup X'_S)$ return a predicate specifying the conditions to take transitions*

Note that we assume each state satisfying its mode's invariant to be a possible initial state. Let $[X_S \to \mathbb{R}]$ be the set of all assignments that map from the

---

[3] The expressiveness depends on the function and relation symbols of the underlying constraint language, we assume basic linear arithmetic, boolean algebra and standard inequalities.

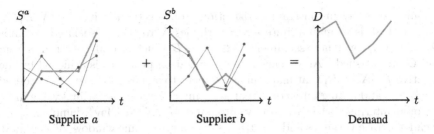

Fig. 3. Resource allocation over multiple time steps can be viewed as selecting the optimal combination of trajectories of two agents to meet a given demand.

variables X_S to $\mathbb{R}$. Let $p \in \Phi(X_S)$ be a set of predicates over X_S. We then write $v \models p$ to denote that an assignment $v \in [X_S \to \mathbb{R}]$ satisfies p.

Definition 2. *The semantics of a supply automaton $SA = (M, X, T, \mathsf{Inv}, \mathsf{Jump})$ are described by an associated transition system $\mathsf{TS}(SA) = (\Sigma, \longrightarrow^*)$ where:*

- *the (infinite) state space $\Sigma \subseteq M \times [X_S \to \mathbb{R}]$ is given by $\Sigma = \{(m, v) \mid v \models \mathsf{Inv}(m)\}$.*
- *the transition relation $\longrightarrow \subseteq \Sigma \times \Sigma$ is defined by $(m, v) \longrightarrow (m', v')$ if there exists a transition $t = (m, m') \in T$ such that $v \cup v' \models \mathsf{Jump}(t)$.*

We assume SA to be completed by τ-transitions, i.e., take the reflexive closure $\longrightarrow^*$ to allow the supply automaton to preserve its state without explicit notations.

A trajectory of a transition system $(\Sigma, \longrightarrow^*)$ is then a sequence $(\sigma_0, \sigma_1, \ldots, \sigma_k)$ such that $\sigma_i \in \Sigma$ for $i \in [0, \ldots, k]$ and $\sigma_i \longrightarrow \sigma_{i+1}$ for $i \in [0, \ldots, k-1]$. We write $\sigma \longrightarrow^{(k)} \sigma'$ to denote that σ' can be reached from σ in exactly k steps. With respect to Example 1, we have to reformulate the trajectories in terms of Σ to verify that $(0, 0, 5, 6)$ is indeed valid (if the supplier is already in the start-up phase). Consider the trajectory $((\mathrm{su}, cd = 1, S = 0) \longrightarrow (\mathrm{su}, cd = 0, S = 0) \longrightarrow (\mathrm{on}, cd = 0, S = 5) \longrightarrow (\mathrm{on}, cd = 0, S = 6))$. By contrast, $(\mathrm{su}, cd = 1, S = 0) \not\longrightarrow (\mathrm{on}, cd = 0, S = 5)$ since the jump condition $cd = 0$ is violated (see Fig. 2).

3.2   Translating Supply Automata into a Synthesized CSOP

In light of supply automata, the resource allocation problem presented in Eq. (1) corresponds to selecting optimal schedules from the set of feasible trajectories for each supplier in an agent organization such that the demand is met by the aggregate supply, as illustrated in Fig. 3. More specifically, we seek assignments of supply and states for future time steps that are consistent with invariants and jump conditions. This question naturally leads to the framework of constraint satisfaction and optimization problems (CSOP) treated in areas of discrete optimization including constraint programming and mathematical programming. We show how to systematically generate such a CSOP based on the syntactical representation of supply automata.

Quite generally, in constraint satisfaction problems (CSP) given by (X, D, C), the main task is to map a finite set of variables X to their associated domains $(D_x)_{x \in X}$ (i.e., construct assignments $v \in [X \to D]$) such that all constraints from a set C are satisfied.[4] An extension to optimization problems is achieved by the objective $f : [X \to D]$ that maps an assignment to a (possibly partially) ordered set $(F, \leq_F)$ and asking for a consistent assignment having an optimal value in F.

For an intermediary i with supply automata $(M^a, X^a, T^a, \mathsf{Inv}^a, \mathsf{Jump}^a)_{a \in \mathrm{sub}(i)}$, we can generate variables and constraints over a given time window $\mathcal{W}$ as follows (we write generated CSOP variables and constraints in typewriter font to avoid confusion with the variables in the supply automata):

Variables. We create the set of decision variables by "flattening" a particular trajectory over the window $\mathcal{W}$. In particular, we need variables $\mathsf{S[a][t]}$ to represent S_t^a for *every* agent $a \in \mathrm{sub}(i)$, a variable $\mathsf{x\_a[t]}$[5] for every *individually required* variable $x \in X^a$, and the mode m^a for every time step $t \in \mathcal{W} \cup \{t_{\mathrm{now}}\}$ written as $\mathsf{m[a][t]}$. The set corresponding to X_S^a for a particular time step t is written as $\mathsf{X_S^a[t]} = \{\mathsf{S[a][t]}\} \cup \{\mathsf{x\_a[t]} \mid x \in X^a\}$. For a predicate $\varphi \in \Phi(X_S^a)$, let $\varphi \langle X_S^a \to \mathsf{X_S^a[t]} \rangle$ denote the substitution of $x \in X_S^a$ with their counterparts in $\mathsf{X_S[a][t]}$, e.g., "$S^a \leq 50$" becomes "$\mathsf{S[a][t]} \leq 50$".

Initial States. The time step t_{now} is needed to consider the current state. For $S_{t_{\mathrm{now}}}^a$, we add the constraint "$\mathsf{S[a][t_{\mathrm{now}}]} = S^a$", and similarly, we require "$\mathsf{x\_a[t_{\mathrm{now}}]} = \sigma^a.x$" for $x \in X^a$ and "$\mathsf{m[a][t_{\mathrm{now}}]} = m$".

State Invariants. For all time steps and all modes, the respective invariant has to hold, i.e., for every $t \in \mathcal{W}$ and $m \in M$, we add the constraint "$\mathsf{m[a][t]} = m \implies \mathsf{Inv}^a(m) \langle X_S^a \to \mathsf{X_S^a[t]} \rangle$"

Transition Constraints. For all transitions $(m, m') \in T$ and every time step $t \in \{t_{\mathrm{now}}\} \cup \mathcal{W} \setminus \{t_{\mathrm{now}} + W\}$, we add the constraint "$\mathsf{m[a][t]} = m \wedge \mathsf{m[a][t+1]} = m' \implies \mathsf{Jump}^a(m) \langle X_S^a \to \mathsf{X_S^a[t]}, X_S^{\prime a} \to \mathsf{X_S^a[t+1]} \rangle$".

The formulation of the CSOP is completed by adding the objective function f. Since the demand does not depend on the agents that should supply it and $\mathsf{S[a][t]}$ is included for every agent, we can use the objective presented in Eq. (1). Summing up, we write SYNTHESIZE$((SA)_{a \in A}, f, (F, \leq_F))$ for the CSOP translated from a set of supply automata.

Example 2. Consider, for instance, two suppliers a and b: a is able to switch between on and off and contributes in the range $[5, 10]$ if on, whereas b only has one mode with its supply ranging in $[4, 8]$; both can maximally ramp up or down 1 unit per time step. Furthermore, upon being turned on, a has to provide its minimal production 5. Assume that both suppliers currently provide their respective minimum, 0 or 4. We consider scheduling $W = 4$ time steps and only care for minimizing discrepancies (i.e., $\alpha_\Gamma = 0$, $\alpha_\Delta = 1$). Constructing the supply automata, we arrive at the following optimization problem that is actually a mixed integer program (presented in pseudocode-OPL):

---

[4] We adopt a functional view, i.e., each constraint $c \in C$ maps an assignment $v \in [X \to D]$ to $\mathbb{B} = \{\mathrm{true}, \mathrm{false}\}$ and consequently, c is satisfied w.r.t. v iff $c(v) = \mathrm{true}$.

[5] Note that the distinction between $\mathsf{x_a}$ and $\mathsf{S[a]}$ emphasizes that S is defined for *every* agent and $\mathsf{x_a}$ is generated *specifically* for a.

$$\texttt{range window} = t_{now} \mathinner{.\,.} t_{now} + 4 \tag{2}$$

$$\texttt{agents} = \{a, b\}$$

$$\texttt{minimize sum(t in window) abs (sum(a in agents) } S[a][t] - D[t])$$

$$\texttt{subject to} \quad \texttt{forall(t in window) } \{$$

$$m[a][t] = \texttt{off} \rightarrow S[a][t] = 0$$

$$m[a][t] = \texttt{on} \rightarrow S[a][t] \geq 5 \text{ and } S[a][t] \leq 10$$

$$m[b][t] = \texttt{on} \rightarrow S[b][t] \geq 4 \text{ and } S[b][t] \leq 8$$

$$m[a][t] = \texttt{on or } m[a][t] = \texttt{off}$$

$$m[b][t] = \texttt{on}$$

$$\}$$

$$\texttt{forall(t in } t_{now} \mathinner{.\,.} t_{now} + 3) \{$$

$$m[a][t] = \texttt{off and } m[a][t+1] = \texttt{on} \rightarrow S[a][t+1] = 5$$

$$m[a][t] = \texttt{on and } m[a][t+1] = \texttt{on} \rightarrow$$

$$\texttt{abs}(S[a][t] - S[a][t+1]) \leq 1$$

$$m[b][t] = \texttt{on and } m[b][t+1] = \texttt{on} \rightarrow$$

$$\texttt{abs}(S[b][t] - S[b][t+1]) \leq 1$$

$$\}$$

$$m[a][t_{now}] = \texttt{off}, \ m[b][t_{now}] = \texttt{on}, \ S[a][t_{now}] = 0, \ S[b][t_{now}] = 4$$

Once the CSOP is synthesized, we can include *soft constraints* [31] regarding individual preferable states and transitions (e.g., "try to schedule at 240 KW" or "never ramp up more than 25 % in 15 min, even if technically possible"). If not all soft constraints can be satisfied simultaneously, suppliers may establish *constraint relationships* [40] prioritizing these goals, i.e., "avoiding strong ramp-up" is more important than "stay at 240 KW". Soft constraints lead to a special kind of codomain for the objective function $(F, \leq_F)$, viz. elements of a c-semiring or valuation structures (partially ordered monoids). Intuitively, an assignment $v \in [X \to D]$ is graded by a set of soft constraints and the overall grading $f(v)$ is found using the multiplication operator of the c-semiring or valuation structure. The precise semantics of these statements along with their integration with Eq. (1) are outside the scope of this work but are described in detail in [25, 38].

4   Abstraction of Composite Control Models

Except for the root of the hierarchy, intermediaries can be regarded as another type of supplier that is also controlled by a superior intermediary. Instead of merely using composed control models on higher hierarchy levels (which would effectively just result in a centralized solution), we introduce some reduction of complexity by an automated abstraction algorithm. Clearly, abstraction may cause errors due to imprecision but leads to a scalable resource allocation scheme as discussed in our evaluation in Sect. 5.1. The task at hand is consequently to formalize the construction of an *abstracted agent model* (AAM), as introduced in Sect. 2, for an intermediary.

Since both AAM and *individual agent models* (IAM) are of the same type to have a superior agent only deal with agent models denoted by supply automata, an AAM also defines possible production ranges and transitions between productions for S_t^i. More abstractly, it describes the set of feasible trajectories of an intermediary's aggregate production. Technically, this set is represented by means of suitable constraints. We are interested in finding the "corners" of the possible supply space spanned by an intermediary as well as "holes", i.e., contributions that cannot be achieved by the set of subordinate agents due to discontinuities, e.g., resulting from discrete on/off transitions. We propose three abstraction algorithms that result in an abstract supply automaton representing the intermediary.

General abstraction (Sect. 4.1) calculates feasible production ranges corresponding to joint modes of the underlying suppliers by considering on/off settings and minimal/maximal supply. *Temporal abstraction* (Sect. 4.2) aims at dismissing provably infeasible schedules due to inertia (e.g., limited ramping rates and start-up times). *Sampling abstraction* (Sect. 4.3) probes a collective's functional relationships such as a mapping from supply to costs and uses a simpler representation on higher levels.

4.1 General Abstraction

The first important abstraction consists of describing the feasible regions of an intermediary based on its subordinate agents. As the contribution of an individual supplier may be discontinuous due to distinct operation modes (e.g., when a supplier shows a strictly positive minimal contribution when *on* but may also not contribute at all when *off*), the production space is discontinuous in general.

Example 3. Consider an intermediary i responsible for two suppliers a and b with their respective possible contributions given by $\{[0,0], [1,4]\}$ and $\{[0,0], [7,10]\}$ where $[0,0]$ indicates that the suppliers might be off. Then every production from $[1,4]$ can be reached by switching supplier a on and supplier b off. Dually, the intermediary can produce $[7,10]$ if only supplier b is running. Engaging both suppliers leads to a combined production interval $[8,14]$ and analogously $[0,0]$ with both being excluded entirely. However, no output in the intervals $(0,1)$ and $(4,7)$ can be provided. Abstraction enters the picture in the sense that it is not relevant whether, e.g., the output $S^i = 8$ is created by setting S^1 to 1 and S^2 to 7 or by S^1 to 0 and S^2 to 8. Multiple concrete configurations collapse to one abstract view, beneficially to the complexity of solving the resource allocation problem. We can thus contract the overlapping intervals $[7,10]$ and $[8,14]$ to yield $[7,14]$. To sum up, the feasible regions of that intermediary are given by $\{[0,0], [1,4], [7,14]\}$ with supply holes $(0,1)$ and $(4,7)$. Hence, 0 and 14 constitute the intermediary's *actual* minimal and maximal production under consideration of underlying on/off-modes.

From this example, we derive a formulation of general abstraction that returns the possible productions of an intermediary. Let $+$ be the standard plus operation in interval arithmetic such that $[x_1, y_1] + [x_2, y_2] = [x_1 + x_2, y_1 + y_2]$ which will be used to calculate the combined production of two suppliers each operating in one particular mode.

These possible contributions of a supplier a are given by a sorted list L^a of non-overlapping intervals. Since a supplier may contribute in *any* of the offered intervals due to varying modes, we need to match any two intervals for combination. For this purpose, we lift the combine operation $+$ to lists and write $\times$ for the combination of two lists. First, we consider the set of all combinations of intervals for two lists:

$$L^a \times L^b := \{[x, y] + [x', y'] \mid [x, y] \in L^a, [x', y'] \in L^b\} \tag{3}$$

For a set of n suppliers A each having k distinct intervals, the set $\prod_{a \in A} L^a$ contains $O(k^n)$ elements. This comes as no surprise since our pivotal application scenario with $L^a = \langle [0, 0], [S^a_{\min}, S^a_{\max}] \rangle$ enumerates the power set of *running* power plants. A brute-force implementation consequently suffers from this exponential behavior. Fortunately, considerable savings can be achieved by composing the overall result from smaller partial results and merging overlapping intervals. In addition, the problem is solvable in linear time if $k = 1$, i.e., if all suppliers have to be *on* and provide in one continuous interval. With respect to our case study, this situation corresponds to so-called "must-run" power plants [48]. Nonetheless, we consider the general case ($k \geq 1$).

For the aforementioned contraction, we recursively define a normalization operation $\downarrow$ which takes a sorted list of intervals (in ascending order of the lower bounds of the intervals) and merges overlapping intervals such that, e.g., $\langle [7, 10], [8, 14] \rangle \downarrow = \langle [7, 14] \rangle$. Concatenation of lists is written as $L_1 \cdot L_2$.

$$\langle \rangle \downarrow = \langle \rangle$$
$$\langle [x, y] \rangle \downarrow = \langle [x, y] \rangle$$
$$(\langle [x_1, y_1], [x_2, y_2] \rangle \cdot L) \downarrow =$$
$$\begin{cases} \langle [x_1, y_1] \rangle \cdot (\langle [x_2, y_2] \cdot L) \downarrow) & \text{if } y_1 < x_2 \\ (\langle [x_1, \max\{y_1, y_2\}] \rangle \cdot L) \downarrow & \text{else} \end{cases} \tag{4}$$

Equipped with this operation, we can implement general abstraction by iteratively calculating the set of possible combinations as shown in Algorithm 1.

As a consequence, the list of feasible regions for an intermediary i is given by $L^i = \text{GENERAL-ABSTRACTION}((L^a)_{a \in \text{sub}(i)})$, also written as $\bigotimes_{a \in \text{sub}(i)} L^a$. Technically, the ability to calculate partial results and normalize them, originates in the associativity of $\times$ (inherited from $+$) and the fact that normalization obtained by $\downarrow$ is only a different representation for the same set of feasible contributions. Regarding supply automata, each interval $j = [x, y] \in L^i$ translates to one mode m_j with $\text{Inv}(m_j) = \text{"}x \leq S^i \leq y\text{"}$. Transitions and jump conditions are added in Sect. 4.3.

Algorithm 1. General Abstraction to find feasible regions

Require: L^a is a finite list of feasible regions for a supplier $a \in A \neq \emptyset$, A is finite
Ensure: L is a list of feasible regions reachable by the suppliers in A

1: **procedure** GENERAL-ABSTRACTION($(L^a)_{a \in A}$)
2:      CHOOSE $a \in A$ ▷ arbitrary first supplier
3:      $L \leftarrow L^a$
4:      **for all** $b \in A \setminus \{a\}$ **do**
5:          $L' \leftarrow \langle \rangle$
6:          **for all** $[x, y] \in L \times L^b$ **do**
7:              INSERT-SORTED($L', [x, y]$)   ▷ sort by x ascendingly
8:          $L \leftarrow L'\!\downarrow$ ▷ normalize
9:      **return** L

4.2 Temporal Abstraction

While general abstraction describes feasible regions of an intermediary, it fails to consider momentary *states*, such as the current productions, that further restrict inertia if some suppliers internally are at peak level or switched off. *Temporal abstraction* thus calculates infeasible ranges of an intermediary i for all time steps $t \in \mathcal{W}$ given the initial state $\sigma^i_{t_{\text{now}}}$ and supply $S^i_{t_{\text{now}}}$. For an intermediary, a possible production level at time step t is composed by the supply offered by its children. Starting from a given initial state, we determine temporally feasible ranges by *both* minimizing and maximizing every child's production respecting jump conditions until time step t and merging the resulting intervals using $\otimes$. Values below or beyond those ranges are guaranteed to be *infeasible* and must therefore not be allocated in the abstract view.

Example 4. Consider the same intermediary i presented in Example 3 responsible for two suppliers a and b specified by $L^a = \langle [0,0], [1,4] \rangle$ and $L^b = \langle [0,0], [7,10] \rangle$. Now assume that a is off at the moment and needs to be off for one more time step before contributing the minimum 1 and b currently producing 8. Both suppliers feature a maximal rate of change of 1 per time step. Provided that we intend to schedule 2 time steps, we analyze possible contributions after each future time step under the present constraints. At $t_{\text{now}}+1$, a may still only be considered with 0 as there is one step more to wait, whereas b's output can be either decreased to 7 or increased to 9 but not be switched off completely. Consequently, the intermediary can only contribute in the range of $[7,9]$ at that point. At $t_{\text{now}}+2$, a can contribute the minimum 1 (or be still off if we chose not to start it) and b can go up to 10 maximally or be already switched off to yield 0 since the minimum of 7 is reachable in the previous step $t_{\text{now}}+1$. Therefore, the intermediary can contribute in $\langle [0,0], [1,1] \rangle \otimes \langle [0,0], [7,10] \rangle = \langle [0,0], [1,1], [7,11] \rangle$. If we incorrectly assumed the full range of general abstraction to be available, wrong allocations, such as $S^i_{t_{\text{now}}+1} \leftarrow 2$ and $S^i_{t_{\text{now}}+2} \leftarrow 3$, would be possible in the abstract view of the intermediary i and need systematic treatment.

The intuition behind temporal abstraction directly stems from Example 4. For each future time step, we perform a *maximization step* as well as a *minimization step* that provides us with the feasible regions of all subordinate agents, which are combined to find abstract boundaries. Technically, for a supplier a modeled by SA^a with its associated transition system $\mathsf{TTS}(SA^a) = (\Sigma, \longrightarrow)$, we look for $S_t^{a,\max} = \max\{S \mid \exists \sigma \in \Sigma : (m_{t_{\mathrm{now}}}^a, \sigma_{t_{\mathrm{now}}}^a) \overset{(t-t_{\mathrm{now}})}{\longrightarrow} \sigma \wedge \sigma.S = S\}$ and dually for the minimization. For the constraints showed in Sect. 3, we can calculate these boundaries by ramping-up and down step-wise but in general this need not hold.[6]

Restricted transitions regarding possible supply are encoded in a supplier's jump conditions Jump^a. Given a state and supply, we assume that all conditions allow to derive functions that present minimal and maximal supply and states in the following time step and call them *inertia functions*. For each mode $m \in M^a$ and $c = \mathsf{Jump}^a(m)$, we require associated functions $c_{\min}$ and $c_{\max}$. We write $\sigma_t^{a,\max}$ and $S_t^{a,\max}$ to denote the states and supplies for supplier a at the maximization step t (dually for the minimization). To illustrate the concept of such inertia functions, consider a fixed rate of change condition $\mathsf{maxChange}$: $|S^a - S'^a| \leq S_\delta^a$. We can derive bounding functions that depend on a state as follows: $\mathsf{maxChange}_{\min}(\sigma^a, S^a) := (\sigma^a, S^a - S_\delta^a)$ and $\mathsf{maxChange}_{\max}(\sigma^a, S^a) := (\sigma^a, S^a + S_\delta^a)$. Similarly, in the case of a start-up condition su, $\mathsf{su}_{\max}(\sigma^a,)$ equals $(\sigma^a\langle m \to \mathsf{on}\rangle, S_{\min}^a)$ if $\sigma^a.cd = 0$ (the remaining "count down" in that state) and $(\sigma^a\langle cd \to cd - 1\rangle, 0)$ if $\sigma^a.cd > 0$. In addition, minimization and maximization steps ought to respect minimal and maximal productions of the respective mode.

Based on these inertia functions, we derive Algorithm 2 to exclude infeasible parts of the search space. For a future time step t, we identify the minimal and maximal contribution of each subordinate a returned by its inertia functions. To obtain the minimal and maximal output of the intermediary in this time step, we combine and merge the resulting intervals with $\otimes$. We write L_t^i to represent the feasible regions of the intermediary i for time step t which corresponds to the combined gray intervals in Fig. 4. These intervals *further* constrain feasible schedules in addition to the general bounds represented by the combined white intervals established by general abstraction.

Temporal abstraction thus adds constraints excluding infeasible regions for specific time steps in the synthesized optimization problem (on higher levels), thereby further reducing the abstracted search space. These constraints are as well only concerned with S_t^i and therefore seamlessly integrate with the AAM found by general abstraction.

$$\forall t \in W : \exists [x, y] \in L_t^i : x \leq S_t^i \leq y \tag{5}$$

In addition to the general boundaries L^i returned by GENERAL-ABSTRACTION(i) (see Algorithm 1) that hold for all time steps, we constrain S_t^i (the supply of i in time step t) to lie in an interval specified by L_t^i. Section 5.4 investigates the effects of temporal abstraction.

---

[6] Consider a supplier with limited fuel resource in-flow such as, e.g., a biogas plant. At a future time step t, the supply could actually be higher if no gas had been spent in the previous steps rather than ramping-up upfront and needing to ramp-down at t due to the lack of fuel.

Algorithm 2. Temporal Abstraction to exclude infeasible ranges

---

Require: i is an intermediary; $\sigma_{t_{\text{now}}}$ contains state t_{now} for all suppliers $a \in \text{sub}(i)$
Ensure: $(L_t^i)_{t \in \mathcal{W}}$ consists of the feasible regions reachable by the intermediary at time step t
1: **procedure** TEMPORAL-ABSTRACTION($i, \sigma_{t_{\text{now}}}$)
2:      intermediaries $\leftarrow \{a$ is an intermediary $\mid a \in \text{sub}(i)\}$
3:      **for all** $j \in$ intermediaries **do**
4:          $(L_t^j)_{t \in \mathcal{W}} \leftarrow$ TEMPORAL-ABSTRACTION($j, \sigma_{t_{\text{now}}}^j$)
5:      $\forall a \in \text{sub}(i) \setminus$ intermediaries $: \sigma_{t_{\text{now}}}^{a,\min}, \sigma_{t_{\text{now}}}^{a,\max} \leftarrow \sigma_{t_{\text{now}}}^a$
6:      **for all** $t \in \mathcal{W}$ **do**
7:          **for all** $a \in \text{sub}(i) \setminus$ intermediaries **do**
8:              $S_t^{a,\min}, \sigma_t^{a,\min} \leftarrow c_{\min}(\sigma_{t-1}^{a,\min}, S_{t-1}^{a,\min})$ $\triangleright c = \mathsf{Jump}^a(\sigma_{t-1}^{a,\min}.m)$
9:              $S_t^{a,\max}, \sigma_t^{a,\max} \leftarrow c_{\max}(\sigma_{t-1}^{a,\max}, S_{t-1}^{a,\max})$ $\triangleright c = \mathsf{Jump}^a(\sigma_{t-1}^{a,\max}.m)$
10:             $L_t^a \leftarrow \{[S_t^{a,\min}, S_t^{a,\max}]\}$
11:         $L_t^i \leftarrow \bigotimes_{a \in \text{sub}(i)} L_t^a$
12:     **return** $(L_t^i)_{t \in \mathcal{W}}$

4.3 Sampling Abstraction

In addition to finding the feasible regions of an intermediary, we are interested in functional relationships between aggregate variables, such as mapping the total production of all subordinate agents to total costs. Similarly, inertia functions for the intermediary depending on current productions ought to be found. Temporal abstraction only goes so far as to exclude *definitely infeasible* productions at time t. It does not restrict the transition between two independently feasible but not consecutively reachable productions for future time steps.

Example 5. In the scenario described in Example 4, we found that $L_{t_{\text{now}}+1}^i = \langle[7,9]\rangle$ and $L_{t_{\text{now}}+2}^i = \langle[0,0],[1,1],[7,11]\rangle$ are feasible. Assume a schedule leaving suppliers a and b at their current state during time step $t_{\text{now}} + 1$ (thus $S_{t_{\text{now}}+1}^i = 8 \in L_{t_{\text{now}}+1}^i$) but asking $S_{t_{\text{now}}+2}^i = 11 \in L_{t_{\text{now}}+2}^i$. Clearly, both scheduled productions are individually attainable with respect to temporal boundaries but ignore the fact that $S_{t_{\text{now}}+2}^i = 11$ is reachable *only if* $S_{t_{\text{now}}+1}^i = 9$. More precisely, b needs to already ramp up to 9 in step $t_{\text{now}} + 1$ to reach 10 in step $t_{\text{now}} + 2$ when a is then able to provide its minimum level 1, leading to a combined production of 11. With regard to Fig. 4, consider a schedule leaving the suppliers at their current output for the time steps $t_{\text{now}} + 1$ and $t_{\text{now}} + 2$ but demanding S_3^i to be maximal. This schedule ignores that, starting from state $\sigma_{t_{\text{now}}}$, only regions in $L_{t_{\text{now}}+1}^i \subset L^i$, i.e., the gray intervals in time step $t_{\text{now}} + 1$ are possible.

Hence, inertia functions for the intermediary mapping a certain aggregate production to possible successor states are of interest. Similarly, an intermediary's cost function is needed to be able to make better allocations on higher levels. We propose to acquire an abstract representation of these functional relationships by *sampling*, i.e., solving several optimization problems that "probe" the collective behavior of an intermediary. Concretely, these sampling problems consist of the constraints given by the composition of agent models and introduce

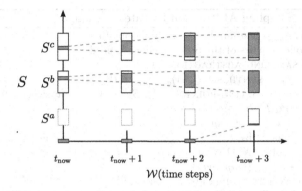

Fig. 4. Temporal abstraction for an intermediary consisting of three suppliers given their current states shown in time step t_{now}. White boxes indicate general bounds, gray areas represent the temporal boundaries at time step t. Supplier a needs two time steps to start up and is then available at its minimal output.

an additional constraint that binds the input variable to some particular value. Regarding inertia functions, as objective, the output variable is minimized or maximized.

For instance, consider an AVPP with 400 being included in its feasible regions obtained by general abstraction. Multiple configurations of its children can lead to these 400, resulting in varying possible increases in production. To find the maximal positive change, we enforce $400 = S^i_{t_{\text{now}}} = \sum_{a \in \text{sub}(i)} S^a_{t_{\text{now}}}$ and ask to maximize $S^i_{t_{\text{now}}+1}$. Assuming that the solution yields 450, we collect the pair $(400, 450)$ as sampled information. Similarly, we find the minimal costs by using the objective to minimize $\kappa(S^i_{t_{\text{now}}}) = \sum_{a \in \text{sub}(i)} \kappa(S^a_{t_{\text{now}}})$ given $S^i_{t_{\text{now}}} = 400$.

Performing this procedure not for one but a set of sampling points evenly distributed over the respective input range is intended to capture the intermediary's characteristics. The resulting input-output pairs can be represented by a suitable approximation method. We currently employ *piecewise linear functions* since they are readily supported by MIP or constraint solvers and have already been applied in model abstraction in simulation engineering [13]. Algorithm 3 illustrates this idea for finding maximal productions given the current state. Since absolute time indices are not needed for this calculations, we may fix S^i_0 and maximize S^i_1. Note that we formulate constraints and the objective syntactically using quotes rather than presenting them semantically in terms of, e.g., their extension.

As of now, the sampling points are selected equidistantly across the full range although more informative points can be selected systematically using techniques borrowed from active learning [39]. The resulting sampling optimization problems (see line 7 in Algorithm 3) are NP-hard in general. Therefore, a robust implementation ought to consider bounding the time spent on optimization by, e.g., setting a time limit. If then no solution is found for a particular input, we give up on it and deal with the next input. Guidance by properties of the

---

Algorithm 3. Sampling Abstraction for rates of change

Require: $(SA)_{a \in \text{sub}(i)}$ is a family of supply automata
Ensure: S_δ^{i+} collects pairs of the positive change speed
1: **procedure** SAMPLING-ABSTRACTION(i, s)
2:     $(X, D, C, f) \leftarrow$ SYNTHESIZE$((SA)_{a \in \text{sub}(i)}, f, (\mathbb{R}, \leq))$ ▷ see Sect. 3.2
3:     $I \leftarrow$ CHOOSE-SAMPLING-POINTS(s, L^i), $S_\delta^{i+} \leftarrow \emptyset$ ▷ select s feasible points
4:     **for all** $inp \in I$ **do**
5:         $C' \leftarrow C \cup \{$"$S_0^i = inp$"$\}$
6:         $f' \leftarrow$ "maximize S_1^i"   ▷ new objective function
7:         $out \leftarrow$ solve (X, D, C', f')
8:         $S_\delta^{i+} \leftarrow S_\delta^{i+} \cup \{(inp, out)\}$
9:     **return** CONVERT-TO-PIECEWISE-LINEAR(S_δ^{i+})

---

function, such as monotonicity ($x \leq y \rightarrow f(x) \leq f(y)$) or extensivity ($x \leq f(x)$), can help to shrink the search space.

It has to be stated that this form of abstraction leads to over-approximations of the actually possible rates of changes and minimal costs. This is due to the fact that among all configurations yielding the input, we select an *extremal* one. For instance, for a certain level of production and possible ramp-up, we select a configuration that offers the most potential (all suppliers far enough below their maximum). In other cases showing the same aggregate supply, some subordinate agents might however be already peaking and cannot offer additional ramp-up. This leads to an over-approximation of the ramp-up. Similarly, when considering a schedule in the abstract view, we assume to achieve the optimal costs for each scheduled production individually (i.e., for each time step), whereas transitions among them might not be technically feasible in the concrete view. More sophisticated abstraction techniques to allow for more robust estimations (e.g., by considering minimal as well as maximal costs for a given supply) have yet to be investigated.

However, regarding our current setting, our evaluation in Sect. 5.2 shows promising results for using the proposed scheme for cost function approximation. Mostly, this is due to the fact that regionally optimal solutions at the lowest level of the hierarchy can improve upon possibly suboptimal decisions made on higher levels. As expected, Table 4 and Fig. 7 show that more sampling points lead to increased accuracy in abstraction observable by better overall costs at the cost of higher running times. An implementation exploiting this principle could start by collecting an initial small set of sampling points to offer a first crude approximation faster while collecting more sampling points in a background process. Finally, the sampled inertia function is used to define the set of transitions and suitable jump conditions for the abstracted supply automaton. Consider two modes m_j, m_l with $j = [x, y]$, $l = [x', y'] \in L^i$ obtained by general abstraction such that $y < x'$. We add transitions (m_j, m_l) if $S_\delta^{i+}(y) \geq x'$ and (m_l, m_j) if $S_\delta^{i-}(x') \leq y$. Furthermore, we add reflexive transitions (m_j, m_j). All of these transitions include the jump condition $S_\delta^{i-}(S^i) \leq S'^i \leq S_\delta^{i+}(S^i)$.

5   Evaluation

To give the presented algorithms empirical grounding, we evaluated the approach on problems taken from decentralized energy management using power plant models built from freely available data [12]. A centralized (optimal) solution planning the outputs of all power plants at once is compared to the regio-central approach using abstraction. These solutions are used for benchmark purposes offering insight in how close to the optimum the regio-central approach gets.

More specifically, we consider biofuel, hydro, and gas plants in the region of Swabia in Bavaria. Nameplate capacities, i.e., maximal productions are drawn from a distribution according to this data. Minimal productions depend on the type of plant and are given as percentage of the nameplate capacity. Similarly, maximal rates of change per minute and costs per kWh are selected based on the type and taken from [21, 26], respectively. To sum up, our data is based on the quantities in Table 2.

This statistical data forms the basis of a generative model that we can use to sample realistic power plants for our simulation. First, we sample the type according to their relative frequency and then draw the maximal production depending on the selected type according to normal distributions ($\mathcal{N}(368.72, 1324.62)$ for biofuel, $\mathcal{N}(264.63, 746.55)$ for hydro, and $\mathcal{N}(275.88, 494.80)$ for gas) within the bounds listed in Table 2. Rates of change and costs are added based on type and maximal production. The demand, i.e., the residual load, to be fulfilled by a set of power plants is based on consumer data of [29] and scaled such that the peak loads map to 110 % of the drawn plants' combined maximal productions – in order to have a representative load test for the system. Regarding Eq. (1), we set α_Δ to a high value[7] to prioritize the goal of meeting the demand but still minimize costs among all demand-satisfying schedules.

Upon drawing a set of power plants, hierarchies are created similar to a B+ tree, i.e., only AVPPs at the lowest level control physical power plants. The hierarchies' shapes, i.e., depth and width, are controlled by restricting the maximal number of physical power plants per AVPP at the leaf level and the maximal number of directly subordinate AVPPs at the inner node levels. First, the "leaf" AVPPs

Table 2. Initially assumed distribution of input data including characteristics [21, 26]. We list minimal and maximal bounds for the maximal production. Standard deviations for cost distribution are given in parentheses.

Type	Rel. Frequency [%]	Max. Power [KW]	Min. Power [%]	Rate of Change [%/min]	Costs [€ / KW]
Biofuel	54	[3.0, 17374.0]	35	6	0.175 (0.014)
Hydro	43	[2.0, 7800.0]	0	50	0.15 (0.017)
Gas	3	[1.0, 2070.0]	20	20	0.0865 (0.004)

---

[7] The value of α_Δ effectively acts as a "market price" since violations would have to be compensated by buying additional energy.

are created by taking a random permutation of the physical plants and picking clusters of the maximal physical plant count. Then, new hierarchy levels in the form of intermediate AVPPs, i.e., inner nodes, are introduced when needed.

We export the synthesized optimization problems (be it regio-central or central ones) as mixed integer programs that are solved by IBM ILOG CPLEX [10] which is indicative for the state-of-the-art in commercial energy management software [35,47]. Given the same power plants and initial states, the problem is solved, unless otherwise stated, for a period of half a day (i.e., 48 time steps in 15 min intervals, being a standard in energy markets) both centrally and regio-centrally — called a *run*. We obtain comparable results by taking care of random seeds and reproducibility. Consequently, all our experiments follow the basic structure:

1. Draw n power plants
2. Load consumer data for half a day (unless otherwise stated, $t = 48$ steps)
3. Solve the resource allocation problem in a hierarchical way
4. Solve the resource allocation problem in a central way
5. Repeat k times (k depends on the experiment)

We investigate parametrizations for the regio-central approach depending on the specific evaluation questions that can be seen as the independent variables in our experiment setup. That includes the number of sampling points to be used, the maximal number of concrete plants per AVPP as well as the maximal number of AVPPs per AVPP to control depth and width of hierarchy structures.

Several dependent variables are measured to compare the performance and are introduced when needed. The most prominent ones are clearly overall costs and runtimes per run as well as per time step which results in either $k \cdot t$ or k data points that are statistically analyzed.

The experiment suite and full source code including an instruction on how to run the experiments can be found online[8] in an attempt to provide replicable research. Each presented experiment was run on a machine having 4 Intel Xeon CPU 3.20 GHz cores and 14.7 GB RAM on a 64 bit Windows 7 OS with 8 GB RAM offered to the Java 7 JVM running the abstractions as well as CPLEX.

All central models used for comparison were solved with a 30 min time limit per time step. When planning for 15 min intervals, a solution must be available much earlier. We still wanted to collect optimal solutions as benchmarks and therefore allowed twice that period for the central solver. We examine questions of interest and present the results of the experiment runs in the following sections.

5.1 Scalability

Does the size of the problem impact the performance in terms of time and quality? We expect that upon reaching a certain number of power plants, the runtime per time step scales better in the hierarchical setting than in a central benchmark. We vary the number of power plants considered and compare the achieved

---

[8] https://github.com/Alexander-Schiendorfer/TCCI-Abstraction.

runtime and costs. Given n power plants, we construct a hierarchy by grouping 35 physical plants at the leaf level as suggested by Sect. 5.3. Inner AVPPs may control up to 10 other AVPPs. By inspecting the runtime behavior of a centralized solution from 5 to 100 plants (see Fig. 6), we found that, although the median runtimes grow linearly with the number of scheduled suppliers, the spread increases strongly and, in particular, outliers showing high runtimes are more probable. We drew 50 times a set of power plants of the respective size and simulated 48 time steps, i.e., half a day.

Table 3 presents the results per time steps aggregated over 2400 data points. For each input size, the difference between monetary costs per step as well as the difference between runtimes per time step were statistically significant using a paired t-test at $\alpha = 0.01$. Figure 5 visualizes this effect, indicating that the central runtimes strictly grow faster than the hierarchical ones. However, the overhead costs incurred by using the hierarchical scheme are in the order of magnitude of 1 % while speed-ups of up to 50 % compared to the central solution could be achieved. Note that we compare the runtime of the centralized approach to the aggregated sequential runtime of all intermediaries in the regio-central approach. Further runtime improvements can be achieved when parallelizing the schedule creation of independent intermediaries, i.e., those that belong to different subtrees in the hierarchy.

In addition to the speed-up per time step, one has to consider the runtime to be invested on the abstraction itself.[9] For a period of 48 time steps, the "fixed costs" (in terms of runtime) for abstraction consist of the computational effort for general and sampling abstraction. With only 50 plants, amortization of the hierarchy is not reached as the mean execution times for the central solution (57.49 s) are far below the hierarchical case (82.57 s). At 100 plants, the central times (158.87 s) begin to exceed the hierarchical case (154.91 s). This development culminates in a difference of 5232.75 s (central) vs. 3210.62 s (hierarchical) when

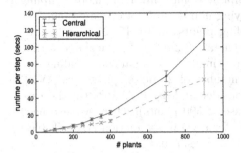

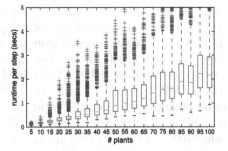

Fig. 5. Runtimes central vs. hierarchical per time step. $\frac{1}{6}$ of the standard deviation is presented for visualization.

Fig. 6. Runtime development of *central* solutions. Only runtimes less than 5 s are plotted. With rising n, the probability of outliers increases drastically.

---

[9] Note that the *temporal* abstraction execution is already included in the runtime per time step for the hierarchical setting.

Table 3. Comparison of central and hierarchical approach over varying numbers of plants. Costs (Γ) and runtimes (T) are presented per time step averaged over 2400 data points from 50 drawn sets of plants each considering 48 time steps. Values in parentheses denote standard deviations.

# plants	T_{centr} (sec)	T_{hier} (sec)	rel.	Γ_{centr} (€)	Γ_{hier} (€)	rel.
50	1.2 (3.42)	1.49 (0.3)	124.18 %	2395.6 (536.14)	2398.21(536.16)	+ 0.11 %
100	3.31 (7.43)	2.82 (0.57)	85.19 %	4777.65 (845.56)	4782.79(845.81)	+ 0.11 %
150	5.27 (6.19)	4.43 (1.32)	83.94 %	7008.41 (1170.96)	7017.81(1169.98)	+ 0.14 %
200	8.09 (8.54)	5.92 (2.45)	73.23 %	9431.81 (1322.07)	9442.39(1321.87)	+ 0.11 %
250	10.67 (5.77)	7.71 (3.62)	72.31 %	11805.68 (1686.42)	11819.61(1686.49)	+ 0.12 %
300	15.2 (9.58)	9.6 (5.37)	63.14 %	14120.96 (1818.34)	14138.42(1818.0)	+ 0.13 %
350	19.15 (13.7)	11.44 (8.16)	59.71 %	16343.07 (2139.34)	16362.32(2139.0)	+ 0.12 %
400	23.2 (13.46)	13.49 (10.16)	58.14 %	18695.82 (2388.73)	18721.15(2390.16)	+ 0.14 %
700	65.68 (37.26)	45.28 (56.43)	68.93 %	32576.8 (3930.08)	32910.47(3906.97)	+ 1.05 %
800	81.19 (45.26)	36.07 (65.84)	44.43 %	37376.75 (4435.96)	37664.91(4431.76)	+ 0.79 %
900	109.02 (75.49)	61.64 (107.57)	56.54 %	41936.44 (4890.82)	42244.31(4835.42)	+ 0.76 %

considering 900 plants. The average relative share of abstraction runtime (excluding temporal abstraction time that belongs to every time step) with respect to the full 48 time step simulation decreases from 13.5 % (50 plants) to 7.8 % (900 plants). Concluding, we see a significant speed-up but, perhaps more importantly, that the abstracted models obtained at reasonable time investments are accurate enough to provide solutions close to the optima.

5.2 Sampling Accuracy

How does the extra running time spent on collecting sampling points pay off in terms of accuracy? Since we expect the abstracted models to increase in accuracy by using more sampling points, considerable improvements should be obtained at the cost of getting these sampling points (involving more sampling optimization problems) upfront. For varying numbers of power plants (50, 200, and 500) and varying numbers of sampling points (5, 10, 15, 25, and 35), we drew 50 times a set of plants from our generative model and calculated 5 time steps both regio-centrally and centrally (optimal) for comparison, totaling in a number of 250 data points. A maximal number of 20 power plants per AVPP and 5 AVPPs per AVPP were set to consider at least 3 AVPPs (in the case of 50 plants) up to 25 AVPPs (in the case of 500 plants), where we also need an additional intermediate layer to allow for several abstraction steps. Table 4 summarizes the results that are visualized in Figs. 7 and 8.

As expected, the runtimes rise with the number of sampling points due to the number of optimization problems that have to be solved in abstraction. Since 250 identical steps (50 draws with 5 time steps each) were solved by the algorithm with parameters, we tested our hypothesis of varying costs per time step using a pairwise t-test at $\alpha = 0.01$. Significantly different costs per step were shown for all combinations of sampling points other than 25 vs. 35 sampling points with 50 plants, 5 vs. 15 and 25 vs. 35 sampling points with 200 plants, and 5 vs. 10

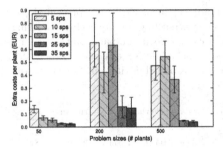

Fig. 7. The influence of sampling points on the accuracy of abstracted models. Extra costs are normalized w.r.t. the number of considered plants. We plot $\frac{1}{4}$ of the std.dev. for visualization purposes.

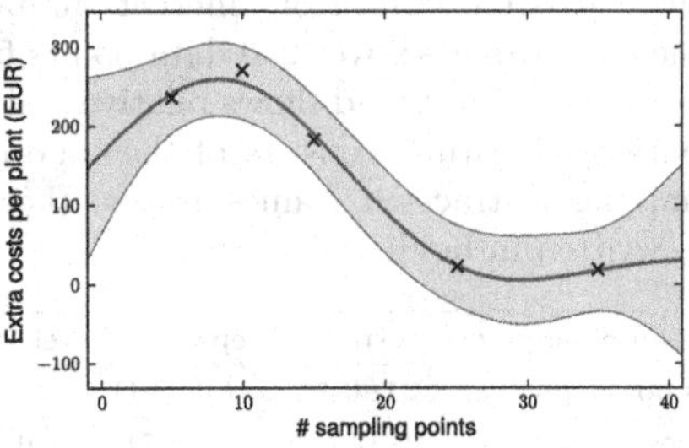

Fig. 8. The influence of the number of sampling points on the overhead costs in the case of 500 plants taken from Table 4. A Gaussian Process regression model is fit to the data to visualize the nonlinear functional relationship.

sampling points with 500 plants. In general, one can see a tendency that very high accuracy is achievable by selecting many points. However, offering more sampling points alone does not guarantee actual improvement due to the fact that the points are just selected equidistantly over the full range and therefore, *informative* points may appear randomly in sets consisting of fewer points but not in the larger one. We observe this behavior, e.g., when comparing 5, 10 and 15 sampling points for 200 plants. By chance, 10 sampling points lead to a significant improvement over 5, whereas 15 points compare unfavorably to 10 and yield accuracy similar to just 5 points. This confirms our findings in [39], emphasizing that more systematic selection is needed. Figure 8 provides more insight into the effect of increasing the number of sampling points on the achieved accuracy. This relationship is clearly nonlinear and one can observe a classical "diminishing returns": adding points to an already saturated set of observed sampling points does provide less benefit than to a comparably smaller set.

5.3 Hierarchy Influence

How does the hierarchy depth and breadth affect the quality and runtime? For this experiment, we varied the maximal number of physical plants per AVPP (ppAVPP) as well as the maximal number of AVPPs per AVPP (avppAVPP) resulting in different system structures for 600, 700, and 800 plants in total. Both parameters lead to broad structures if they are large and deep structures if they are small, thus provoking more abstraction steps.

Table 5 lists the results for different numbers of physical plants as well as ppAVPP/avppAVPP-combinations. Using a paired t-test at $\alpha = 0.01$, we compared all ppAVPP/avppAVPP-combinations in terms of runtime and costs, showing significant differences in most cases. Setting ppAVPP to a small value (5 or 15) leads to fragmented structures where accuracy and efficiency can be wasted as the centralized solver still has capacities for higher numbers of plants, confirming

Table 4. Comparison of different sampling point settings. Costs are presented per time step averaged over 250 data points from 50 drawn sets of plants each considering 5 time steps. Overhead shows relative and absolute extra costs compared to the optimal solutions. Runtime consists of the time needed to schedule 5 steps and overhead for sampling abstraction. Values in parentheses denote standard deviations. Optimal costs are written in bold.

# sampling points	costs per step	rel. overhead	abs. overhead	runtime
50 power plants:	**2180.71 €** (453.91)			
5 sps	2187.73 € (453.57)	+ 0.32 % (0.29)	+ 7.03 € (2.37)	27.29 s (2.82)
10 sps	2184.31 € (453.87)	+ 0.17 % (0.16)	+ 3.6 € (1.79)	41.42 s (4.16)
15 sps	2183.45 € (453.72)	+ 0.13 % (0.15)	+ 2.75 € (1.77)	52.81 s (2.52)
25 sps	2182.05 € (454.01)	+ 0.06 % (0.06)	+ 1.34 € (1.1)	78.11 s (2.92)
35 sps	2181.9 € (453.86)	+ 0.05 % (0.07)	+ 1.19 € (1.15)	106.86 s (3.67)
200 power plants:	**8587.1 €** (938.36)			
5 sps	8717.41 € (1014.05)	+ 1.52 % (1.53)	+ 130.31 € (12.28)	107.05 s (3.61)
10 sps	8671.07 € (975.07)	+ 0.98 % (1.36)	+ 83.97 € (11.2)	172.89 s (33.44)
15 sps	8713.46 € (995.13)	+ 1.47 % (2.22)	+ 126.36 € (14.02)	225.95 s (43.85)
25 sps	8618.24 € (938.57)	+ 0.36 % (0.83)	+ 31.14 € (8.27)	329.84 s (20.76)
35 sps	8616.31 € (952.83)	+ 0.34 % (0.7)	+ 29.21 € (8.14)	460.12 s (51.72)
500 power plants:	**21157.72 €** (1853.51)			
5 sps	21393.53 € (1885.9)	+ 1.11 % (1.08)	+ 235.81 € (14.97)	244.0 s (11.28)
10 sps	21428.24 € (1868.18)	+ 1.28 % (1.16)	+ 270.52 € (15.51)	357.59 s (10.71)
15 sps	21339.92 € (1870.11)	+ 0.86 % (0.99)	+ 182.2 € (14.37)	470.32 s (18.24)
35 sps	21176.55 € (1853.2)	+ 0.09 % (0.09)	+ 18.83 € (4.28)	934.63 s (26.98)

our findings in Fig. 6. The largest value for ppAVPP, 35, led to the best runtime and cost efficiency in all considered cases which is why it was used as setting for Sect. 5.1. Interestingly, the tradeoff between cost and runtime performance seems to be regulated by the depth of the hierarchy; deep structures (avppAVPP = 5) led to the best runtime performance at slightly higher costs due to more abstractions. In the case of 600 plants, however, the differences in cost with 35 plants per AVPP over all AVPP per AVPP settings was not significant. Similarly, the difference in cost between 35/5 and 35/15 was not significant for 700 plants. With 900 plants, however, the flat structure with 15 AVPPs per AVPP showed significantly different costs compared to 5 and 10 AVPPs per AVPP. Concluding, our results suggest that AVPPs at the leaf level need to schedule an adequately number of power plants, whereas the speed-up benefits of having a deep structure controlled by small inner nodes seem to outweigh the (in most cases insignificant) additional costs. A more thorough analysis including probabilistic models of the system structures and more ppAVPP/avppAVPP-combinations is planned.

5.4 Temporal Abstraction for Inertia

Is temporal abstraction required for accurate abstracted models? We want to investigate the effects of temporal abstraction on the achieved accuracy. Therefore, we analyzed a small setting consisting of 50 power plants that are either

Table 5. Influence of the hierarchy on costs and runtimes per time step. Costs are presented per time step averaged over 250 data points from 50 drawn sets of plants each considering 5 time steps. Left column contains runtimes in seconds; right column contains costs in Euros; values in parentheses denote standard deviations. Best configurations per number of plants are written in bold.

plants per AVPP	5 AVPPs per AVPP		10 AVPPs per AVPP		15 AVPPs per AVPP	
600 plants:						
5 plants per AVPP	34.27	25895.19	43.92	25840.34	34.43	25842.65
	(1.63)	(2110.12)	(2.16)	(2107.79)	(1.4)	(2106.21)
15 plants per AVPP	33.66	25780.93	31.07	25765.1	31.64	25784.81
	(1.42)	(2137.22)	(1.39)	(2120.32)	(1.28)	(2122.12)
35 plants per AVPP	**25.1**	25729.94	27.96	25712.88	28.1	**25712.09**
	(2.29)	(2080.1)	(9.93)	(2108.26)	(6.89)	(2106.86)
700 plants:						
5 plants per AVPP	39.65	30109.09	50.27	30048.11	40.02	30045.2
	(1.61)	(2444.44)	(1.67)	(2422.91)	(1.56)	(2418.81)
15 plants per AVPP	39.51	29992.62	35.91	29966.94	36.51	29972.93
	(3.9)	(2449.15)	(1.57)	(2447.9)	(1.4)	(2429.5)
35 plants per AVPP	**28.77**	29897.34	30.18	29934.82	38.46	**29896.27**
	(1.73)	(2431.65)	(4.19)	(2445.54)	(18.2)	(2430.87)
800 plants:						
5 plants per AVPP	56.47	34542.19	59.43	34456.06	45.8	34458.11
	(1.98)	(2747.09)	(2.64)	(2737.62)	(1.85)	(2737.22)
15 plants per AVPP	41.93	34502.89	41.36	34310.24	42.25	34377.36
	(2.19)	(2738.86)	(1.49)	(2742.44)	(1.62)	(2750.6)
35 plants per AVPP	**32.81**	34377.9	34.52	34383.45	41.5	**34289.73**
	(2.18)	(2722.67)	(2.28)	(2711.98)	(14.74)	(2722.37)

grouped in 10 AVPPs of 5 power plants or in 5 AVPPs of 10 power plants that are managed by the root AVPP. Both cases show a hierarchy of depth 2 and therefore need one step of abstraction. Furthermore, the length of one time step has been reduced from 15 min to 1 min to consider increased effects of inertia. Errors are defined as differences between schedules assigned to intermediaries and their possible redistribution to their own subordinates.

As Table 6 shows, the error between assigned schedules to an intermediary and actually achievable schedules, i.e., the difference between its own subordinates' overall supply and assigned share of the residual load, can be reduced by employing temporal abstraction. This difference in errors is significant using a pairwise t-test at $\alpha = 0.01$ for both 5 as well as 10 plants per AVPP. Also, the error roughly doubles with 5 plants since more abstractions are performed and more possible deviations are introduced. Interestingly, the tighter rates of change induced by the shorter scheduling time steps makes the problem

Table 6. Influence of temporal abstraction. Costs are presented per time step averaged over 2400 data points from 50 drawn sets of plants each considering 48 time steps. Runtime is measured per time step (excluding fixed times for general and sampling abstraction); error denotes the overall discrepancy between supply and demand. Values in parentheses denote standard deviations.

plants per AVPP	T_{centr}	T_{hier}	Rel.	Error
without temporal abstraction:				
5 plants	12.25 (127.69)	4.36 (1.92)	35.55 %	3.55 (55.06)
10 plants	10.67 (116.23)	4.1 (0.56)	38.37 %	1.37 (33.9)
with temporal abstraction:				
5 plants	14.33 (141.99)	3.89 (0.58)	27.15 %	**0.26** (6.25)
10 plants	15.01 (143.04)	4.41 (2.49)	29.38 %	**0.01** (0.09)

significantly harder for CPLEX. This setting closely resembles the one used in [41], explaining the proportionally higher speed-up of about 70 % compared to Sect. 5.1 with about 50 %. Therefore, the proposed abstraction techniques can prove particularly useful when short intervals are considered to react timely on updated prognoses, slow power plants have to be scheduled, or schedules *should* not utilize the full technical potential of a power plant's rate of change, i.e., are restricted by soft constraints.

6   Related Work

Our proposed approach draws inspiration from many well-established concepts from diverse areas of computer science, including abstract interpretation, model abstraction, decomposition, and finite state machines.

The methodology of using abstracted models that are then refined on lower levels is closely related to abstract interpretation [9] used in program verification or approximation techniques used in the analysis of hybrid systems [17]. The latter technique analyzes minimal and maximal derivatives $\dot{x}$ of a state variable x (e.g., supply) w.r.t. time given a model in the form of differential equations and produces so-called rectangular hybrid automata that have constant upper and lower bounds. Hence, in one mode, x can perform continuous time transitions according to $\dot{x}_{\min} \leq \dot{x} \leq \dot{x}_{\max}$, similar to our considerations in Sects. 4.2 and 4.3. Merging output variables shared by a composition of agents (the aggregate supply) as we do in general abstraction is not considered. However, approximations obtained by their technique could serve as supply automata.

Abstractions have been treated from many different angles and in a variety of fields. First attempts toward automatic "model abstraction" to make pre-defined abstractions of complex physical system models obsolete were made in [46]. From an AI perspective, [15] was an attempt to systematically describe different varieties of abstraction and their formal properties, especially with regard to deduction and representation. A similar systematization was done in [13] where a taxonomy of common model abstraction techniques from the perspective of the simulation engineering community is provided. Crucially, [13] states that abstraction techniques "must maintain the validity of the simulation results with respect to the question being addressed [...]" which holds true in the abstraction of composite control models as well. Another approach is to differentiate structural and behavioral abstraction [28]. The authors demonstrate abstraction methods and, e.g., use neural networks to get a behavioral abstraction of subcomponents that were given as state machines.

Abstraction methods for task and resource allocation problems are presented in [8]. One of the techniques, summarization, has similarities to our general abstraction since it combines time intervals in much the same way that we combine production intervals (cf. Sect. 4.1). The other technique, generalization, uses a taxonomy of the domain concepts used in the constraints to find more general formulations, e.g., to express that a certain operating room belongs to a certain unit within a hospital. Instead of assigning a surgeons to concrete operating rooms, they are assigned to units, making the problem smaller. We do not employ a similar technique. The paper argues for the inclusion of user decisions in the abstraction process to ensure that abstraction errors are avoided.

An approach to extract constraint models from an architectural description of the system for use in runtime optimization is presented in [16]. A model of the instantiation of an abstract architecture is used to derive constraints that describe the correct configurations of a system. The utility function takes into account the utility of the configurations as well as the utility and cost of the reconfiguration and the decision making itself. The resulting problem can either be solved with a MIP approach or by pseudo-boolean optimization (PBO) in order to find the best configuration of the system given the circumstances. Interestingly, the evaluation results suggest that the PBO method that intuitively should yield more accurate results failed to deliver and an approach based on MIP is sufficient. Similar to our work, the optimization problems are constructed at runtime, but no abstraction is attempted. Since the systems regarded do not follow a hierarchical structure, the problem solved has a very different structure.

The problem we address has similarities with lot size problems, especially those with minimal order quantities [33]. These problems address the optimal size of orders in cases where the machines producing them have a setup cost as well as a unit production cost and there is the possibility to store an inventory of items for later delivery. Minimal order quantities give a lower bound of the production per time step that is similar to the one for power generators: either a machine is switched off completely or it produces at least a minimal number of units. Formulations with a capacity constraint also limit the maximum inventory [27]. Such a limit can be used to model the maximum capacity of storage power plants in a power grid. However, the formulations from operations research do not address two important issues in energy management: change rates and loading times. The proposed models assume that it is possible to switch from maximum production (usually bounded by the cumulative demand) to zero production within one time step. Likewise, the inventory can take up as many units as produced within a single time step. Power generators, however, have ramp up times that limit the change in their production from one time step to another. Power storage facilities, on the other hand, have loading rates that limit the energy that can be stored per time step. These issues are reflected in the supply automata we present in Sect. 3.

7   Conclusions and Future Work

In this paper, we addressed a resource allocation problem that is very relevant to energy management systems. The variant we consider shows *minimal* in addition to maximal supply capacities, discontinuity by on/off switches, and inert suppliers. We have motivated the need for self-organizing hierarchical structures due to scalability concerns and the large search space of "good hierarchies" in terms of runtime and monetary efficiency. To allow for self-organization, we need to separate individual properties, such as the start-up behavior of suppliers, from overall goals: meeting the demand cost-efficiently. In the process, we suggested supply automata to model individual, heterogeneous supplies as suitable formalism for common problems in the power systems literature that can be automatically translated into synthesized optimization problems and provided an example.

Based on supply automata, we discussed techniques to calculate abstractions of the composite models of a sub-system, i.e., intermediaries. Our evaluation shows that the hierarchical setting achieves a quality within 1 % of the optimal solutions at about 30 % to 50 % of the runtime by mere decomposition, i.e., without a parallel execution of the individual sub-systems. The runtimes of the hierarchical approach for given input sizes showed more favorable growth than a centralized benchmark using CPLEX. While the presented abstraction methods have been successfully applied to a market-based scheduling approach [2], we hope that the presented algorithms and models also generalize well to other problem domains.

The paper focused on the modeling and optimization aspects of the overall hierarchical, multi-agent approach sketched in Sect. 1. Uncertainties arising from deviations of predictions are dealt with by means of robust optimization methods using trust-based scenarios as described in [3]. Cooperative, i.e., truthful behavior of the involved agents is incentivized using techniques inspired by mechanism design [2]. It remains to be analyzed if a truthful mechanism for revealing the supply automata (the "types" in the language of algorithmic mechanism design) is achievable and/or desirable.

Our experiments also revealed further directions of research: given that the quality of abstraction depends strongly on the number and location of sampling points (see Sect. 5.2), we plan to investigate better choices for a guided selection strategy for sampling points using active learning or response surfaces. First tests show significant potential for improvement [39]. Furthermore, Sect. 5.3 highlighted the influence of the hierarchy on solution quality and runtime. Current techniques [43] rely on local rules triggering reconfigurations. Assuming patterns in the relationship of, e.g., input size, type composition and runtime, we hope to learn more about the hardness of subproblems given certain features using techniques from empirical algorithmics such as model-based algorithm configuration [22].

References

1. Abouelela, M., El-Darieby, M.: Multidomain hierarchical resource allocation for grid applications. J. Electr. Comput. Eng. **2012**, 8 (2012)
2. Anders, G., Schiendorfer, A., Siefert, F., Steghöfer, J.P., Reif, W.: Cooperative resource allocation in open systems of systems. ACM Trans. Auton. Adapt. Syst. **10**, 11 (2015)
3. Anders, G., Schiendorfer, A., Steghöfer, J.P., Reif, W.: Robust Scheduling in a Self-Organizing Hierarchy of Autonomous Virtual Power Plants. In: Stechele, W., Wild, T. (eds.) Proceedings of the 2nd International Workshop Self-optimisation in Organic and Autonomic Computing Systems (SAOS 2014), pp. 1–8 (2014)
4. Anders, G., Siefert, F., Reif, W.: A particle swarm optimizer for solving the set partitioning problem in the presence of partitioning constraints. In: Proceedings of the 7th International Conference on Agents and Artificial Intelligence (ICAART 2015), pp. 151–163. SciTePress (2015)
5. Arroyo, J.M., Conejo, A.J.: Modeling of start-up and shut-down power trajectories of thermal units. IEEE Trans. Power Syst. **19**(3), 1562–1568 (2004)
6. Boudjadar, A., David, A., Kim, J.H., Larsen, K.G., Mikučionis, M., Nyman, U., Skou, A.: Hierarchical scheduling framework based on compositional analysis using uppaal. In: Fiadeiro, J.L., Liu, Z., Xue, J. (eds.) FACS 2013. LNCS, vol. 8348, pp. 61–78. Springer, Heidelberg (2014)
7. Chevaleyre, Y., et al.: Issues in multiagent resource allocation. Informatica **30**(1), 3–31 (2006)
8. Choueiry, B.Y., Faltings, B., Noubir, G.: Abstraction methods for resource allocation. Technical report, Swiss Federal Institute of Technology in Lausanne (EPFL) (1994)
9. Cousot, P.: Abstract interpretation. ACM Comput. Surv. **28**(2), 324–328 (1996)
10. CPLEX: IBM ILOG CPLEX Optimizer, Dec 2013. online Resource. http://www-01.ibm.com/software/commerce/optimization/cplex-optimizer/. Accessed December 2013
11. Dash, R.K., Vytelingum, P., Rogers, A., David, E., Jennings, N.R.: Market-based task allocation mechanisms for limited-capacity suppliers. IEEE Trans. Syst. Man Cybern. Part A: Syst. Hum. **37**(3), 391–405 (2007)
12. Deutsche Gesellschaft für Sonnenenergie e.V.: Energymap, Dec 2013. http://www.energymap.info/. Accessed December 2013
13. Frantz, F.: A taxonomy of model abstraction techniques. In: Proceedings of the Winter Simulation Conference 1995, pp. 1413–1420 (1995)
14. Frey, S., Diaconescu, A., Menga, D., Demeure, I.: A generic holonic control architecture for heterogeneous multi-scale and multi-objective smart micro-grids. ACM Trans. Auton. Adapt. Syst (2015)
15. Giunchiglia, F., Walsh, T.: A theory of abstraction. Artif. Intell. **57**(2), 323–389 (1992)
16. Götz, S., Wilke, C., Richly, S., Piechnick, C., Püschel, G., Assmann, U.: Model-driven self-optimization using integer linear programming and pseudo-boolean optimization. In: International Conference on Adaptive and Self-Adaptive Systems and Applications (ADAPTIVE 2013), pp. 55–64 (2013)

17. Henzinger, T.A., Ho, P.H., Wong-Toi, H.: Algorithmic analysis of nonlinear hybrid systems. IEEE Trans. Autom. Control **43**(4), 540–554 (1998)
18. Heuck, K., Dettmann, K.D., Schulz, D.: Elektrische Energieversorgung. Vieweg+Teubner (2010). (in German)
19. Hladik, P.E., Cambazard, H., Déplanche, A.M., Jussien, N.: Solving a real-time allocation problem with constraint programming. J. Syst. Softw. **81**(1), 132–149 (2008)
20. Horling, B., Lesser, V.: A survey of multi-agent organizational paradigms. Knowl. Eng. Rev. **19**(4), 281–316 (2004)
21. Hundt, M., Barth, R., Sun, N., Wissel, S., Voß, A.: Verträglichkeit von erneuer-baren Energien und Kernenergie im Erzeugungsportfolio. Technisch-ökonomische Aspekte. Studie des Instituts für Energiewirtschaft und rationelle Energieanwen-dung (IER) im Auftrag der E. ON Energie AG. Stuttgart (2009). (in German)
22. Hutter, F., Hoos, H.H., Leyton-Brown, K.: Sequential model-based optimization for general algorithm configuration. In: Coello, C.A.C. (ed.) LION 2011. LNCS, vol. 6683, pp. 507–523. Springer, Heidelberg (2011)
23. Jarass, L., Obermair, G.M.: Welchen Netzumbau erfordert die Energiewende?: Unter Berücksichtigung des Netzentwicklungsplans Strom 2012. MV-Wissenschaft, Monsenstein und Vannerdat (2012). (in German)
24. Karl, J.: Dezentrale Energiesysteme: Neue Technologien im liberalisierten Energiemarkt. Oldenbourg (2012). (in German)
25. Knapp, A., Schiendorfer, A., Reif, W.: Quality over quantity in soft constraints. In: Proceedings of the 26th International Conference on Tools with Artificial Intel-ligence (ICTAI 2014), pp. 453–460 (2014)
26. Kost, C., Mayer, J.N., Thomsen, J., Hartmann, N., Senkpiel, C., Philipps, S., Nold, S., Lude, S., Saad, N., Schlegl, T.: Levelized cost of electricity- renewable energy technologies. Techno-Economic Assessment of Energy Technologies, Fraun-hofer ISE (2013)
27. Lee, C.Y.: Inventory replenishment model: lot sizing versus just-in-time delivery. Oper. Res. Lett. **32**(6), 581–590 (2004)
28. Lee, K., Fishwick, P.A.: A methodology for dynamic model abstraction. SCS Tran. Simul. **13**(4), 217–229 (1996)
29. LEW Verteilnetz GmbH: LEW Netzdaten, December 2013. http://www.lew-verteilnetz.de/. Accessed December 2013
30. Mayer, J.N., Kreifels, N., Burger, B.: Kohleverstromung zu Zeiten niedriger Börsenstrompreise, August 2013. http://www.ise.fraunhofer.de/de/downloads/pdf-files/aktuelles/kohleverstromung-zu-zeiten-niedriger-boersenstrompreise.pdf/view
31. Meseguer, P., Rossi, F., Schiex, T.: Soft constraints. In: Rossi, F., van Beek, P., Walsh, T. (eds.) Handbook of Constraint Programming, Chap. 9. Elsevier (2006)
32. Nethercote, N., Stuckey, P.J., Becket, R., Brand, S., Duck, G.J., Tack, G.R.: MiniZ-inc: towards a standard CP modelling language. In: Bessière, C. (ed.) CP 2007. LNCS, vol. 4741, pp. 529–543. Springer, Heidelberg (2007)
33. Okhrin, I., Richter, K.: The linear dynamic lot size problem with minimum order quantity. Int. J. Prod. Econ. **133**(2), 688–693 (2011). towards High Performance Manufacturing
34. Padhy, N.P.: Unit commitment- a bibliographical survey. IEEE Trans. Power Syst. **19**(2), 1196–1205 (2004)
35. PLEXOS Integrated Energy Model, January 2014. http://energyexemplar.com/software/plexos-desktop-edition/. Accessed 09 January 2014

36. Ramchurn, S.D., Vytelingum, P., Rogers, A., Jennings, N.R.: Putting the 'Smarts' into the smart grid: a grand challenge for artificial intelligence. Commun. ACM **55**(4), 86–97 (2012)
37. Santos, C., Zhu, X., Crowder, H.: A mathematical optimization approach for resource allocation in large scale data centers. Technical report, HPL-2002-64, HP Labs, March 2002
38. Schiendorfer, A., Knapp, A., Steghöfer, J.-P., Anders, G., Siefert, F., Reif, W.: Partial valuation structures for qualitative soft constraints. In: Nicola, R., Hennicker, R. (eds.) Wirsing Festschrift. LNCS, vol. 8950, pp. 115–133. Springer, Heidelberg (2015)
39. Schiendorfer, A., Lassner, C., Anders, G., Lienhart, R., Reif, W.: Active learning for abstract models of collectives. In: Proceedings of the 3rd International Workshop Self-optimisation in Organic and Autonomic Computing Systems (SAOS15), March 2015
40. Schiendorfer, A., Steghöfer, J.P., Knapp, A., Nafz, F., Reif, W.: Constraint relationships for soft constraints. In: Bramer, M., Petridis, M. (eds.) Proceedings of the 33rd SGAI International Conference on Innovative Techniques and Applications of Artificial Intelligence (AI 2013), pp. 241–255. Springer (2013)
41. Schiendorfer, A., Steghöfer, J.P., Reif, W.: Synthesis and abstraction of constraint models for hierarchical resource allocation problems. In: Proceedings of the 6th International Conference on Agents and Artificial Intelligence (ICAART'14), vol. 2, pp. 15–27. SciTePress (2014)
42. Schiendorfer, A., Steghöfer, J.P., Reif, W.: Synthesised constraint models for distributed energy management. In: Proceedings of the 3rd International Workshop Smart Energy Networks & Multi-Agent Systems (SEN-MAS 2014), pp. 1529–1538 (2014)
43. Steghöfer, J.P., Behrmann, P., Anders, G., Siefert, F., Reif, W.: HiSPADA: Self-organising hierarchies for large-scale multi-agent systems. In: Proceedings of the IARIA International Conference on Autonomic and Autonomous Systems (ICAS) 2013, IARIA (2013)
44. Ströhle, P., Gerding, E.H., de Weerdt, M.M., Stein, S., Robu, V.: Online mechanism design for scheduling non-preemptive jobs under uncertain supply and demand. In: Proceedings of the International Conference on Autonomous Agents and Multi-agent Systems (AAMAS 2014), pp. 437–444. International Foundation for Autonomous Agents and Multiagent Systems, Richland (2014)
45. Wang, C., Shahidehpour, S.: Effects of ramp-rate limits on unit commitment and economic dispatch. IEEE Trans. Power Syst. **8**(3), 1341–1350 (1993)
46. Weld, D.S., Addanki, S.: Task-driven model abstraction. In: 4th International Workshop on Qualitative Physics, pp. 16–30 (1990)
47. Werner, T.: DEMS - The Decentralized Energy Management System (2013). http://w3.siemens.com/smartgrid/global/en/smart-grid-world/experts-talk/pages/dems.aspx. Accessed 07 January 2014
48. Yingvivatanapong, C., Lee, W.J., Liu, E.: Multi-area power generation dispatch in competitive markets. IEEE Trans. Power Syst. **23**(1), 196–203 (2008)

Shape Recognition Through Tactile Contour Tracing

A Simulation Study

André Frank Krause[1,2]([⊠]), Nalin Harischandra[1,2], and Volker Dürr[1,2]

[1] Department of Biological Cybernetics, Bielefeld University, Bielefeld, Germany
{andre_frank.krause,nalin.harischandra,volker.duerr}@uni-bielefeld.de
[2] Cognitive Interaction Technology - Centre of Excellence (CITEC),
Bielefeld University, Bielefeld, Germany

Abstract. We present Contour-net, a bio-inspired model for tactile contour-tracing driven by an Hopf oscillator. By controlling the rhythmic movements of a simulated insect-like feeler, the model executes both wide searching and local sampling movements. Contour-tracing is achieved by means of contact-induced phase-forwarding of the oscillator. To classify the shape of an object, collected contact events can be directly fed into machine learning algorithms with minimal pre-processing (scaling). Three types of classifiers were evaluated, the best one being a Support Vector Machine. The likelihood of correct classification steadily increases with the number of collected contacts, enabling an incremental classification during sampling. Given a sufficiently large training data set, tactile shape recognition can be achieved in a position-, orientation- and size-invariant manner. The suitability for robotic applications is discussed.

Keywords: Tactile sensor · Contour-tracing · Shape recognition · Artificial neural network

1 Introduction

The tactile sense enables humans and animals to actively perceive their immediate surrounding through direct physical contacts with an object (Prescott and Dürr, 2015). In contrast to vision, direct tactile sampling of an object allows to "feel" object properties like surface texture, chemical properties, temperature, compliance and humidity, that are hard to obtain otherwise (Lederman and Klatzky, 2009). The sense of touch is independent of light conditions, and works equally well night and day. Moreover, the direct contact with an external object may yield reliable distance information (Patanè et al., 2012). Therefore, the tactile sense could potentially play an equally important role in engineering as it does throughout the animal kingdom, especially in nocturnal species.

Several insect species, for example the honey bee (*Apis mellifera*), the American cockroach (*Periplaneta americana*) and the Indian stick insect (*Carausius morosus*) have become important model organisms for the study of the sense

© Springer-Verlag Berlin Heidelberg 2015
N.T. Nguyen et al. (Eds.): TCCI XX, LNCS 9420, pp. 54–77, 2015.
DOI: 10.1007/978-3-319-27543-7_3

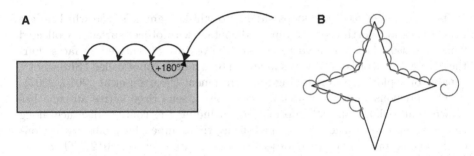

Fig. 1. A: The core idea of Contour-net. Each contact induces a phase forwarding of a circular movement by 180°. Combined with velocity control of a tactile probe, the resulting change in movement velocity and direction causes the probe to "bounce off" the surface at every contact event, resulting in a successive scan along the object's shape. **B**: 2D simulation of contact-triggered contour-tracing. *Black*: star-shaped object. *Blue*: trajectory of the antennal tip. *Red dots*: contact locations (Color figure online).

of touch. Insects carry a pair of antennae that are densely covered with sensory hairs of different modalities (Staudacher et al., 2005). Honeybees, for example, show a high concentration of tactile hairs at their antennal tip (Esslen and Kaissling, 1976). Active tactile scanning of surfaces allows them to discriminate the micro texture of flowers (Kevan and Lane, 1985) or artificial gratings (Erber et al., 1998).

An important constraint of the insect tactile system is that antennae essentially are one-dimensional structures that are incapable of providing a two-dimensional image "at a glance". Instead, antennae need to be moved actively in order to sample information from different locations within their working-range. Active tactile sensing is of particular relevance in near-range exploration. Many insects actively use their antennae for obstacle localization, orientation behaviour, pattern recognition, and even for communication (Staudacher et al., 2005). Similarly, mammals like cats or rats use active whisker movements to detect and scan objects in the vicinity of the body.

Insect antennae and mammal whiskers have inspired robotic research in the area of tactile sensors. Early work by Kaneko et al. (1998) describes an artificial antenna using a flexible beam capable of detecting 3D contact locations and surface properties. Russell and Wijaya (2003) applied an array of whiskers that passively scan over an object to recognize its shape using advanced preprocessing of contact data and decision trees. In Solomon and Hartmann (2006), robotic whisker arrays were used to generate 3D spatial representations of the environment and extract object shapes. Related work done by Kim and Möller (2007) used a vertical whisker array to detect the vertical shape and curvature of objects. In Sullivan et al. (2012), a bio-mimetic tactile sensing system called the "BIOTACT Sensor" is presented, featuring a conical, three-dimensional array of actuated whiskers where each whisker can detect deflections using sensors at its base. Using feedback loops similar to whisked animals, the whiskers are moved back-and-forth to make repeated contact with object surfaces. Collected whisker sensor data was used to classify the texture of whisked surfaces. Further, the active whisker array was mounted on a mobile robot called "Shrewbot", capable of exploring and mapping its environment (Pearson et al., 2013, 2007).

Our previous work on a stick-insect-inspired, sensorised active antenna has shown that reliable spatial information, including the contact distance along the antenna, can be acquired by measuring the damped harmonic oscillations of the elastic probe caused by object contact (Patanè et al., 2012). Therefore, it should be suitable not only to detect obstacles, but also to scan their shape. This can enable autonomous robots to make navigation decisions, for example if a sampled object is a large wall or a climbable stair case. This paper explores in simulation, to what extent sampling using an active artificial antenna can yield reliable shape information about objects.

The model is based on behavioural observations made with stick insects (*Carausius morosus*). Stick insects continuously and rhythmically move their antennae during locomotion (Dürr et al., 2001; Krause et al., 2013b) and adapt the antennal movement pattern during turning (Dürr and Ebeling, 2005), after loss of foothold (Dürr et al., 2001) and during tactile probing of external objects (Krause and Dürr, 2012; Schütz and Dürr, 2011). Upon antennal contact with an obstacle, stick insects modulate both the frequency and the amplitude of the rhythmic antennal movement, affecting both antennal joints in a context-dependent manner (Schütz and Dürr, 2011; Krause and Dürr, 2012). For example, when climbing a square obstacle, they show a contact-induced switch in behaviour from a broad, almost elliptical searching pattern to a local sampling pattern with higher frequency and lower amplitude (Krause and Dürr, 2012). This switch to a local sampling strategy may be interpreted as an effort to gather more detailed spatial information close to previous touch locations, effectively leading to the sampling of an obstacle's contour.

A fundamental concept explaining such rhythmic movements in vertebrates and invertebrates is the central pattern generator (CPG, for a review see Ijspeert, 2008). The CPG activity is often modulated by sensory input, proprioceptive input and descending signals from higher brain centres. Antennal movements in stick insects are assumed to be driven by CPGs, because rhythmic activity can be evoked pharmacologically (Krause et al., 2013b). A recent computational model successfully applies a set of coupled central oscillators to simulate the experimentally observed antennal movements (Harischandra et al., 2015). Using a similar type of oscillator, the present paper presents a simple but effective model that can trace the contour of an object using an actively movable tactile probe. The oscillator activity is modulated by a single, binary sensor signal, telling it whether or not the tactile probe is in contact with the obstacle.

Furthermore, we show that the sampled contact events can be used for 2D and 3D object recognition based on tactile shape classification, even with short sampling sequences comprising less than 20 contacts. Three classifiers with different learning algorithms were applied to the shape classification task. Section 3.3

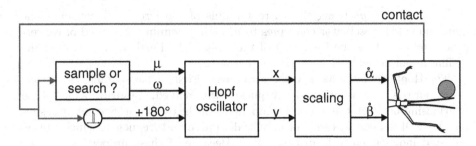

Fig. 2. Block diagram of Contour-net. The binary contact signal (red) determines the frequency and amplitude of the Hopf oscillator. It further triggers an instantaneous phase forwarding of the oscillator by 180°, inducing a movement away from the object surface. The discrete impulse block guarantees that the oscillator phase is forwarded only once. The output of the oscillator is then properly scaled and may drive either a simulated or real robot antenna using velocity control of both antennal joints (Color figure online).

evaluates the differences in classification accuracy and computing time among the three classifiers. We show that the likelihood of correct classification steadily increases with the number of collected contact events if the training data is structured in a specific way. This enables an incremental classification during a sampling trial without requiring a complete scan of the object, making the system suitable for mobile robotic applications.

2   Contour-Net

The contour-tracing model presented here was coined *Contour-net*, hinting at its possible integration into existing, modular architectures for simulation and control of hexapod walking (*Walknet*, e.g., Cruse et al., 1998; Dürr et al., 2004; Schilling et al., 2013) and ant-inspired navigation (*Navinet*, Hoinville et al., 2012). The model captures the essential characteristics of antennal tactile sampling in stick insects: rhythmic searching movements by means of an antenna with two revolute joints, strong and persistent inter-joint coupling, and a contact-triggered switch from a searching to a sampling mode. Such rhythmic movements can be obtained as limit cycles of nonlinear dynamical systems, typically systems of coupled nonlinear oscillators. In our case, the ideal choice is a *Hopf oscillator*, because the two state variables of the oscillator exhibit a fixed phase coupling and can directly drive the two joints of an antenna.

2.1   Hopf Oscillator

The Hopf oscillator is a dynamical system defined in Cartesian space by the following differential equations:

$$\dot{x} = \gamma \left(\mu^2 - r^2\right) x - \omega y$$
$$\dot{y} = \gamma \left(\mu^2 - r^2\right) y + \omega x$$

with $r = \sqrt{x^2 + y^2}$ being the current radius of the circular movement. The amplitude of the oscillator converges to μ, with γ defining the speed of convergence and ω setting the frequency of the limit cycle. Further, the phase of the oscillator can be set with $x = \cos(\varphi)$, $y = \sin(\varphi)$.

The Hopf oscillator has several advantages: First, it has a stable limit cycle behaviour (i.e., perturbations decay quickly) with a fixed, non-drifting, 90° phase relationship between the x and y components of the oscillator. Second, it is simple and well defined in terms of amplitude, phase and frequency, which can be adjusted independently from each other. Because of these properties, smooth online modulation of trajectories can be achieved through changing parameters of the system at run-time. Moreover, these properties help in entrainment of the CPG rhythm through sensory feedback, e.g., when being coupled with a mechanical system (Righetti and Ijspeert, 2006). Hopf CPG's have been applied successfully to biped (Buchli et al., 2005; Righetti and Ijspeert, 2006) and quadruped (Brambilla et al., 2006; Ijspeert et al., 2007) locomotion.

2.2   Contact-Triggered Contour-Tracing

The basic idea of the contour-tracing model, as illustrated in Fig. 1A, is that each contact event triggers a phase shift in the cyclic sampling movement of the antenna. Ideally, this phase shift should cause the antenna to "bounce off" the object after each contact. Direct control of the antennal position with the Hopf oscillator would require memorizing the location of the centre of oscillation, and shifting it along the object surface. The need for a memory structure can be avoided if the Hopf oscillator output is used to set the antennal velocity rather than position. Figure 1B shows the "scan path" (in analogy to eye-tracking scan paths) along a star-shaped 2D-object. The velocity commands applied to the antennal joints, α and β are given by:

$$\Delta\alpha = s_1 x$$
$$\Delta\beta = s_2 y$$

where s_1 and s_2 are scaling factors, setting the velocity of the antenna. Setting these factors to distinct values leads to ellipsoid trajectories as found in stick insect antennal movements (Krause and Dürr, 2004).

The 2D simulation shown in Fig. 1B was implemented in Matlab with contact locations being calculated using the Geom2D toolbox (Legland, 2012). The simulated antenna and the Hopf oscillator equations are iterated using first-order forward Euler integration with a fixed time step Δt.

After a contact is detected by the antenna, the amplitude μ and the frequency ω are immediately switched from a large-amplitude, low-frequency "searching mode" to a low-amplitude, high-frequency "sampling mode". Once a certain time span, $T_{sampling}$, has passed without encountering a further contact event, the parameters are switched back to the "searching mode" pattern.

A 180° phase shift can be easily implemented by negating the Hopf oscillator state variables: $x_{t+1} = -x_t$ and $y_{t+1} = -y_t$. Because of the discrete-time nature

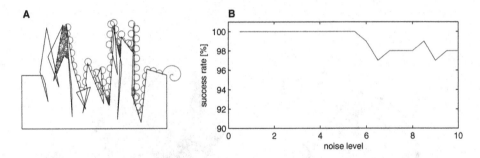

Fig. 3. Robustness evaluation of the contour-tracing algorithm. **A**: Extreme example, in which the antenna got trapped in a cavity of the random contour. **B**: Success rate of contour-tracing in percentage. $N = 100$ trials. Please note that the y-axis values range from 90 % to 100 %. The success rate is above 96 % for all noise levels, showing the high robustness of the contour tracing algorithm.

of numerical simulations, contacts with an obstacle may last longer than a single time step. Therefore, the 180° phase shift should be applied only once for each contact event. In the block diagram in Fig. 2, this is indicated by the discrete pulse block.

2.3   Robustness Evaluation

The robustness of the contour-tracing algorithm was evaluated by scanning randomly generated rough contours consisting of linear segments of random orientation and length. See Fig. 3A for a sample contour. The contour was structured as a 10 units long, initially straight line with 40 uniform segments. The locations (x_i, y_i) of the nodes connecting the segments were then randomized by $x_i = x_i + 0.15 * rand(-r, r)$ and $y_i = y_i + rand(-r, r)$ where $rand(-r, r)$ generates uniformly distributed random numbers between $-r$ and r. The task for the algorithm was to scan completely the contour from the right to the left outermost side without getting stuck. $N = 100$ trials were performed for each noise level $r = 0.5, 1, .., 10$. In almost all trials, the rough surface could be scanned completely from start to end. Figure 3A shows one of the rare cases, where the contour-tracing algorithm got trapped in a cavity. Figure 3B shows the overall success rate, i.e. the percentage of trials that completely scanned the rough surface.

2.4   3D Simulation

The contour-tracing algorithm also worked reliably in 3D space with virtually no change to the algorithm. As listed in Table 1, the only difference between the 2D and 3D models concerns the radius of the movement. Figure 4 shows a kinematic antennal model with orthogonal joint axes, similar to a cardan joint but with a short segment between the axes. The 3D model successfully probed several object types, including a torus or an octahedron. Contact events were determined by use of the Matlab package Geom3D, including the contact distance along the tactile probe.

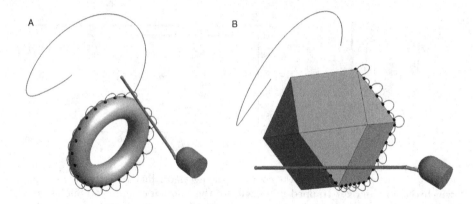

Fig. 4. 3D simulation of contact-triggered contour-tracing. Grey: objects. Green: insect head with left antenna. Blue: trajectory of the antennal tip. Red dots: contact locations. **A**: Torus, **B**: Octahedron (Color figure online).

Table 1. Parameters of the 2D and 3D models.

Parameters	2D-Sim	3D-Sim
γ	4.0	4.0
$\mu_{searching}$	1.0	0.5
$\mu_{sampling}$	0.2	0.1
$\omega_{searching}$	1.0	1.0
$\omega_{sampling}$	2.0	2.0
Δt	0.02	0.02
$s_\alpha = s_\beta$	1.0	0.5

3 Tactile Shape Recognition

An application scenario for Contour-net is tactile shape recognition of sampled objects. Each contact with the surface can deliver direct information about the contact distance, the current joint angles at contact time and intrinsic values like the state of the Hopf oscillator. Collecting these values should allow a discrimination of object shapes based on previous learned examples. Here, we apply a plain but large, three-layered feed-forward Artificial Neural Network (ANN) to solve the classification task. Previous results have shown that multi-layer ANNs can classify hand-written digits and temporal eye-tracking data with minimal pre-processing, only (Krause et al., 2013a, 2011). In our present application, we use scaled contact event data as input values to the ANN without any further pre-processing.

With increasing dimensionality of data sets, the size of the input space grows exponentially. At the same time, the density of data samples within the input space decreases exponentially. Therefore, high-dimensional data sets require a

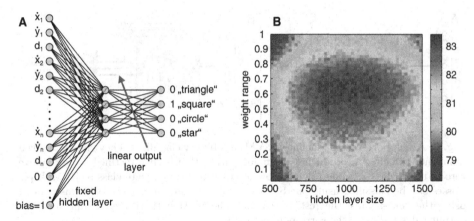

Fig. 5. A: General network structure used for shape recognition. A three-layered feed-forward network with a large hidden layer was used. Data components ($\dot{x}$, $\dot{y}$ from Hopf oscillator, contact distance d along antenna) collected during a contour-sampling were serialized, rescaled to the range -1 to 1 and fed into the network. Shorter contour-scans were zero-padded. Hidden layer weight values were initialized randomly and only the output layer was trained (see text). Class labels use a "one-out-of-n" coding. **B**: Classification accuracy (color coded) depending on weight range and hidden layer size. Used training data set: 3D shapes, trained with 1000 samples per shape and 20 repetitions per sample. The optimal weight range value is independent from the hidden layer size (Color figure online).

large number of training samples to achieve good machine learning results. This is known as the "curse of dimensionality" problem, where each data point can essentially become an outlier (Bellman, 1961). This gives rise to two practical problems: First, the large training data set has to be collected in the first place, which in a practical application can be difficult or extremely time-consuming. Second, large data sets require long training times with common neural network training algorithms. Using No-Prop-*fast* (Krause et al., 2013a), a special case of an Extreme Learning Machine (Huang et al., 2006), the second problem can be circumvented, and almost interactive training times can be achieved. This allows the generation of parameter tuning curves with many repetitions of the learning process, even when using standard personal computers. In order to relate the performance of ANN classifiers to other standard solutions, we also trained a Support Vector Machine (SVM) to solve the same task.

3.1 2D Shape Classification

To collect a large training data set, four different 2D-shapes (triangle, square, star, circle) were traced using Contour-net. The shape size was scaled between 100 % and 200 %, shapes were rotated randomly between 0° and 360°, and the initial contact location was also randomized. Figure 7 shows a small sample from the 2D-data set. A fixed number of contact events was selected such that all

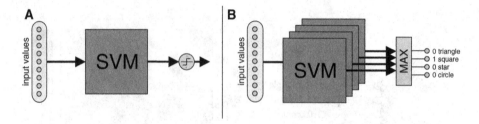

Fig. 6. A: Support Vector Machines (SVM) are binary classifiers. They classify a given input into two classes, producing an analog output that is then rectified. Values above zero indicate class 1, while values below or equal to zero indicate class 0. **B**: Multiple classes can be classified using one SVM per class. In our case, each SVM was trained using the "one-against-the rest" - scheme. The SVM with the maximum "analog" output value indicates the corresponding class.

shapes in all sizes were 'encircled' at least once, hence completely traced. 400 random samples with 35 contact events were collected for each one of four shapes, resulting in a training data set with 1600 samples. Each contact event yielded two "measurements" as input to the classifier: the $\dot{x}$ and $\dot{y}$ velocity components of the Hopf oscillator. As a substitute for contact distance (see Sect. 4) the third input to the network was the shape size, because in 3D the contact distance along the feeler in combination with the Hopf oscillator values can give hints about the size of an object. The third input channel was scaled to a range from $[-1, 1]$, while the first two input channels always range from -1 to 1. The input data to the classifier had 105 dimensions. A "one-out-of-n" coding scheme was used for the output. Hence, the No-Prop network had 105 input units and four linear output units. The weights of hidden layer units in No-Prop networks are fixed (Widrow et al., 2013) and were initialized with uniformly distributed random numbers. The hidden layer units use a sigmoid activation function $(tanh)$. The number of hidden units and their random weight range are described in Sects. 3.2 and 3.3. Figure 5A shows the general neural network structure.

3.2   Classification Performance

25 randomly initialized networks were trained on the data set, and performance was evaluated using ten-fold cross-validation (Kohavi et al., 1995). K-fold cross validation splits the data set into k blocks, where $k - 1$ blocks are used as the training data set and the remaining block is used for validation (Refaeilzadeh et al., 2009). The training is repeated k times, where each block is once the validation set. This gives a more accurate estimation of the performance of the learning algorithm. Figure 8C shows how the classification accuracy depends on the hidden layer size. As shown in Krause et al. (2013a), essentially two components influence the performance: the hidden layer size and the hidden layer weight range. The optimal weight range of hidden layer units does not depend on the number of hidden layer units but on the number of inputs (Fig. 5B).

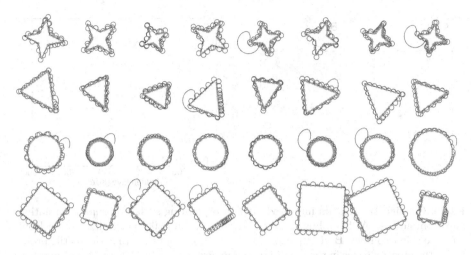

Fig. 7. Sample collection of the four different shapes used for the classification task. The size and orientation of the shapes as well as the initial contact location was randomized.

Its best range was estimated to be -0.5 to 0.5 for both the 2D and 3D training data set (see Sect. 4). Figure 8C (blue curve) shows that a rotation-invariant shape classification could be achieved in about 75 % of the cases. The discrimination rate peaks around 600 hidden units and decreases after that. The low spread of performance values indicates that the algorithm has not to deal with local minima. This is a favourable feature of the No-Prop method, see also Widrow et al. (2013). Compared to gradient-based neural network training, the hidden layer size seems to be fairly large. This normally indicates over-fitting, resulting in poor generalization. Huang et al. (2012); Huang (2014) discusses in detail why such neural networks with large but random hidden layers can have good generalization properties.

Contour-tracing trajectories were found to be fairly variable, depending on angle of attack of the tactile probe relative to the surface. Figure 8A shows an example where the probe arrived at a shallow angle to the surface. The subsequent, contact-induced 180° phase shift of the oscillator caused the probe to leave the surface in a suboptimal, backward direction. The resulting pattern alternated between small and large arcs.

Approaching a surface in the normal direction, i.e., with the angle of attack being 90°, not only avoids potential slip (see Kaneko et al., 1995) but also improves the regularity of the scan path. To test the effect of such improved regularity of the scan path on classification performance, we estimated the surface normal for each contact and adjusted the phase-forwarding to the Hopf oscillator accordingly, thus stabilising the outgoing angle after contact around 90°. In the simulation, the surface normal was estimated by a local, 360° subsampling of the surface around the contact position. From the resulting list of intersections with the object, an average surface normal vector was calculated,

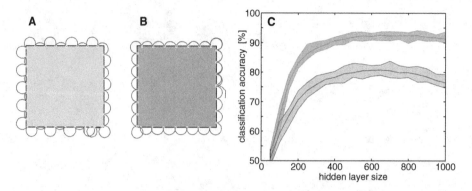

Fig. 8. Classification results improved after incorporating surface-normal information into Contour-net. **A**: Sampling trajectory using a fixed 180° phase shift of the Hopf oscillator after contact. **B**: The phase of the oscillator is set such that the tactile probe leaves the object surface in the normal direction after a contact. The resulting sampling trajectory is more regular. **C**: Classification accuracy using a fixed phase shift (blue), and variable phase shift depending on the surface normal measured at contact time (red). A training data set with four shapes as shown in Fig. 7 with 400 samples for each of the four shapes was used. $N = 25$ repetitions. Solid line: mean value; shaded area: min. and max. values (Color figure online).

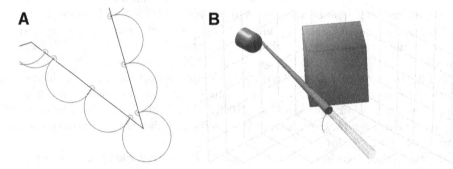

Fig. 9. Estimation of the local surface normal. The immediate surroundings of the current contact is sampled and an average direction vector is calculated from collected intersections with the object. The inverse direction of this vector gives an approximation of the surface normal. **A**: 2D contour tracing simulation. **B**: 3D contour tracing simulation.

similar to the mean resultant vector in circular statistics (see Fig. 9). Figure 8B shows an example of an improved sampling trajectory after consideration of the surface normal. The red curve in Fig. 8C, shows a marked improvement of the classification accuracy. Since there was no slip in our simulation, this result implies that a more regular scan path can improve shape classification performance from 75 % to 90 %. The advantage for the classifier is a lower spread of contact parameters along linear contour sections, and a more pronounced change in contact parameters at the corners of a contour.

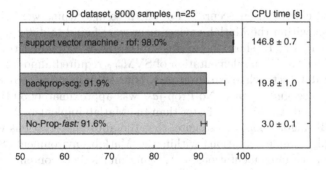

Fig. 10. Classification accuracy and training duration (computing times) of three algorithms compared. First row: No-Prop-*fast* (n_{hidden} = 3000), second row: scaled conjugate gradients backpropagation (n_{hidden} = 64), third row: Gaussian kernel support vector machine. Bars show median, whiskers show min and max values, CPU times are given as mean±SD.

3.3 Performance Evaluation

Shape recognition performance was tested for three classifier variants: the ANN trained with the No-Prop-*fast* method (Krause et al., 2013a), the ANN trained with scaled conjugate gradient backpropagation (BP$_{scg}$, Møller 1993), i.e., the default pattern classification algorithm of the Matlab 2012a neural network toolbox, and a Gaussian kernel Support Vector Machine (SVM$_{rbf}$), as implemented in the Matlab 2012a statistics toolbox. Figure 6 depicts the SVM classification model.

A separate SVM$_{rbf}$ was trained for each shape (one-against-all classification, see Fig. 6B), using the Matlab function *svmtrain()* with the default Sequential Minimal Optimization method and parameters set to *sigma* = 4.2 and *boxconstraint* = 10. The Matlab parameters translate to the standard SVM parameters *gamma* = 0.0283 and penalty parameter C = 10.

The performance of BP$_{scg}$ was tested using a three-layered feed-forward network with 64 hidden units, created using the Matlab function *patternnet()* and trained with all parameters at their default values (maximum number of epochs = 1000; minimum performance gradient = 1e-6; maximum validation failures = 6; *sigma* = 5.0e-5; *lambda* = 5.0e-7). The number of units in the hidden layer was optimized for maximum validation performance using grid search.

The three different algorithms were evaluated using ten-fold cross-validation and 25 repetitions with randomized samples and randomly initialized networks. Figure 10 shows classification accuracies and CPU times of these algorithms, applied to the 3D shape data set with 9000 samples. CPU time measurements were performed on a Dell Precision T3500 Workstation (Intel Xeon W3530 CPU at 2.8 GHz with 8 Mbyte second level cache) running Windows 7. SVM$_{rbf}$ achieved the highest median classification accuracy (98 %). The classification accuracy of BP$_{scg}$ was lower (92 %) with high spread, but individual networks achieved up to 96 %. No-Prop-*fast* came third with 91 %, but with a much smaller variance than BP$_{scg}$.

While the classification accuracies of all three algorithms were above 90 %, training times within the Matlab environment were found to differ significantly. Values given in Fig. 10 show the average duration of individual training runs ($n = 250$). The Matlab implementation of SVM_{rbf} required almost 147 seconds to learn the data set, while No-Prop-*fast* required only three seconds. Hence, for the shape recognition task, No-Prop-*fast* was approximately 50 times faster than SVM_{rbf} and still six times faster than the Matlab implementation of one of the fastest BP-algorithms. Please note, that the above observations were based on "CPU-only" implementations within the Matlab environment. Specialized high-performance implementations, e.g., using multi-GPU computing, and different parameter constellations might result in different overall training times and performance trajectories.

3.4   Incremental Shape Recognition

In contrast to the shape classification tasks used so far, contact sequences acquired by a robotic tactile system may strongly vary in length. Indeed, for application on a hardware system, it would be desirable to achieve robust shape classification for contour traces as brief as possible. Therefore, it was important to test Contour-net and the subsequent shape classifier on data sets with realistically low contact numbers. However, on-line training on the robot poses a severe problem: Classification based on machine-learning algorithms generally requires a high-quality training data set in order to achieve an acceptable classification performance with good generalization capabilities. This is particularly important for data sets with high-dimensional input patterns. With regard to the alternative machine learning algorithms applied above, a drawback of ANN classifiers is that the training data set needs to be fairly large (see Fig. 18A). This becomes even more relevant when considering that, on a real robot, collecting a reasonably sized training data set is likely to be far more costly than generating a data set in a computer simulation. In the following, we limit the assessment of incremental shape recognition to SVM classifiers. The SVM classifier not only gave the highest classification accuracy (Fig. 10) but a preliminary test suggested that a SVM classifier yields more robust results when trained on small data sets. Moreover, for envisaged robot experiments, the relatively long training times of SVM's are outweighted by the time required for collecting a large training data set that would be suitable for an ANN classifier.

Our first goal was to generate an artificial training data set that resembled real data acquired by the robotic system as closely as possible. We reasoned that, provided the match was good enough, it could be possible to train the classifier using an artificial data set and then later test the classifier on a small validation data set acquired by the robot hardware. That way, the amount of robot-acquired training data could be reduced considerably. The most likely object sizes as "seen" by a robot antenna such as the one proposed by Hoinville et al. (2014) were estimated to range between 25° and 40° in a spherical coordinate system. In order to determine the minimum size of the training data set required for acceptable classification performance, the sampling amplitude of Contour-net

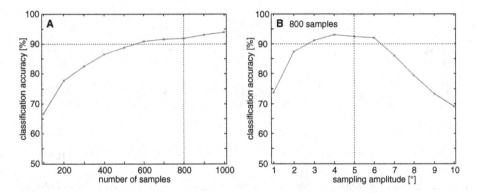

Fig. 11. Effect of sampling amplitude and size of training data set. **A:** An SVM was trained using training data set sizes ranging from 100 to 1000 samples. The classification accuracy was then evaluated using an independent validation data set with $n = 5000$ samples. **B:** The sampling amplitude influences the classification accuracy. For a data set with objects ranging in apparent size from 25° to 40°, the classification accuracy peaks around 4°. Both the training data set and the independent validation set were generated from scratch for each sampling amplitude.

was covaried. In contrast to the ten-fold cross-validation used so far, here we generated a separate, large validation data set consisting of 5000 samples. This was because, firstly, it is relatively easy to generate a large data set in simulation, and secondly, a large validation set increases the reliability of the classification performance estimate.

Figure 11 shows how the classification performance depends on (i) the size of the training data set and (ii) the sampling amplitude, using support vector machines (SVM). Figure 12 gives examples how the contour tracing is affected by the sampling amplitude, i.e., parameter μ of the Hopf oscillator. In both figures, error bands are missing, owing to the deterministic nature of the SVM for a given, fixed pair of training and validation data sets. Figure 11B shows how the contour tracing depends on the appropriate ratio of object size and sampling amplitude. For a data set comprising 800 samples acquired with a sampling amplitude of 5°, classification accuracy exceeded 90 %.

As a consequence, a fixed sampling amplitude of 5° was used in all subsequent simulation experiments. As an option for further refinement of the approach, the sampling amplitude could be adapted to the perceived size of the object, e.g., by including a preliminary visual estimate of object size.

In a next step, we wondered whether it would be possible to train a classifier using Contour-net such that it could give a preliminary estimate of the object shape with less than 15 contact events, and update this estimate with every subsequent contact encountered during a continuous scan. In other words, we were aiming for an incremental shape classification that could provide an early "first guess", leaving the option for a "change of mind" with subsequent contacts, while progressively increasing the reliability of shape classification.

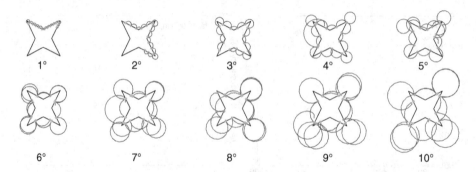

Fig. 12. Effect of sampling amplitude. Example contour traces of a star-shaped object with a perceived size of 25°, using sampling amplitudes between 1° to 10°. All traces show 15 contacts per shape. Larger sampling amplitudes tend to under-sample the object, while small sampling amplitudes do not result in a complete scan of the object.

In a first attempt to test this, we used an SVM trained on the previous data set comprising 800 random shapes, traced with 15 contact events each. In this case, the system was bad at classifying contours with less than 10 contacts (see Fig. 13, blue curve). Therefore, a new training data set was generated, where, in addition to the size and orientation of the shapes, the number of contacts was randomized, too, varying from one to 15 contact events. In order to use a single classifier for input patterns with different contact event numbers, the input pattern always contained 15×3 components, with $n \times 3$ being measurements from n contact events, and $(15 - n) \times 3$ being set to zero. In total, contour traces from 200 random shapes were collected, amounting to 3000 samples. The validation data set comprised 5000 samples, as before. Figure 13, red curve, shows that classification performance for short traces improved considerably, given this training data set. Contour traces with as few as six contact events yielded a classification accuracy of 80 %, with increasing accuracy for each successive contact. This suggests that our method of incremental shape recognition may be suitable for online shape classification. Furthermore, the reliability of shape classification may be estimated by analyzing the analog SVM output values, as illustrated by Fig. 14: The larger the contrast between the maximum output value and the rest, the more reliable is the shape estimate.

3.5   Response to Unknown Shapes: Shape Morphing

Finally, we wanted to know how the classifier would respond to other, unknown object shapes. To find out, we tested the performance to novel shapes with gradually varying similarity to the shapes used during training. For this, the classifier was presented with shapes that were morphed from a disc into a triangle. 25 shapes with random size and orientation were tested at each "morph stage". Figure 15 shows how the ratio of classification counts per shape changes with the morph stage. In most cases, morphs were classified as either a disc or a

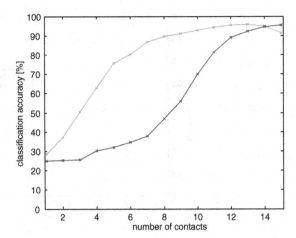

Fig. 13. Successive classification. Classification accuracy depends on the number of collected contact events. Blue trace: classification accuracy of a SVM trained on the default data set with 15 contact events from 800 random shapes. In this case the increment of accuracy was very low for short contour traces of the validation data set. Red trace: If the training data set comprised contour traces of variable length (one to 15 contacts from 200 random shapes), the increment of accuracy increased strongly right from the first contact, and online shape classification was very good for traces as short as six contacts (Color figure online).

triangle. With increasing "triangularity", some shapes were detected as squares or stars, but the likelihood of such mis-assignments was low. This means that the classifier responds to unknown shapes in an expected and predictable manner: it tries to find the closest match to a learned shape and does not suddenly confuse the unknown shape with an unlikely choice like a star. To illustrate this further, Fig. 16 shows the output values of the four SVMs. As a disc shape with fixed size and orientation was slowly morphed into a triangle, the output of the SVM "circle" was largest until a morph stage of approximately 50 %. After that, values of the SVMs "circle" and "triangle" were very similar for some time, as if the classifier had difficulties in deciding between two alternatives. As the morph stage approached a triangle, the output of the SVM "triangle" becomes largest. At no point in the sequence did the outputs of the other two SVMs reach similarly large values, as if the classifier was always sure that the shape was neither a star nor a square.

4   3D Shape Classification

Classification performance was also tested using 3D objects. A major difference between the 2D and 3D simulation was that in 3D the contact distance along the antenna could be measured. Accordingly, the object size entry to the classifier was replaced by contact distance. Moreover, the relative location of the objects could be varied.

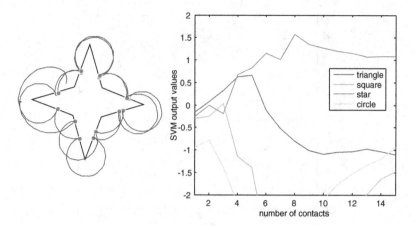

Fig. 14. Example of successive classification. Analog output values of the four SVM's, where each SVM was trained to discern one class of shapes from the rest. In this case, a contour trace of a star was used (left). In the beginning, the star was confused with a triangle, but after 6 contacts the correct classification dominated, with increasing reliability.

Three different objects were used: a soccer ball, a cube and a tetrahedron (Fig. 17). The objects were scaled between 100 % and 150 %, placed at random positions with $(x, y, z) = [20..30, -5..5, -5..5]$ units in front of the antenna, and rotated randomly around all axes in the full range of $360°$. The initial antennal location around the object was also randomized. Three data sets with 400, 1000 and 3000 random samples per shape with 35 contact events per shape were collected. Given three measurements per contact event ($\dot{x}$, $\dot{y}$, and contact distance), the No-Prop neural network had 105 inputs and three outputs. All input values were scaled to the range $[-1, 1]$, identical to Sect. 3.1.

Figure 18 shows the classification accuracy for the three data sets (ten-fold cross validation). The increased complexity of the task required a very large data set to achieve a classification rate of 90 %. Due to the random rotation around all axes, the supervised learning algorithm had to be trained on a sufficiently large number of contour projections of the objects. Contact distance information improved classification performance only slightly, by 5 %, as shown by comparing the performance with and without this third input component (Fig. 18). Hence, the contact distance along the antenna is not essential and does not need to be very precise.

5 Discussion

Contour-net is a bio-inspired method for tactile contour-tracing. It draws its robustness from exploiting the limit cycle of a Hopf oscillator, maintaining a stable 90 degree phase shift between the two output variables. It is bio-inspired

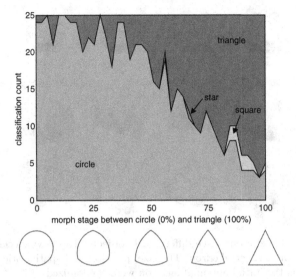

Fig. 15. Response to novel shapes. An SVM trained on four 2D shapes was presented with shapes morphed between a triangle and a circle. The shapes had random size and orientation. 25 examples were presented per morph stage. The classification count decreases rather smoothly, with very few instances in which a shape was detected that was not used to generate the morph.

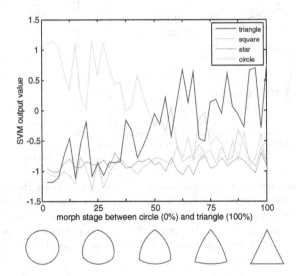

Fig. 16. Analog output values of the SVMs for each class, given the contour trace of one shape with fixed size and orientation but varying morph stage.

in that it captures characteristic aspects of tactile searching and sampling behaviour observed in its natural paragon, the stick insect antenna (Dürr et al., 2001). For example, it implements a contact-induced switch from a large-amplitude,

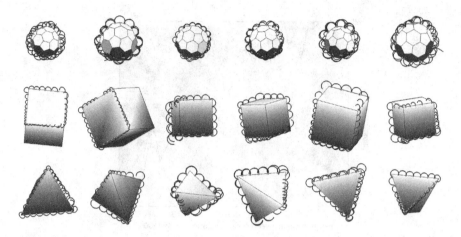

Fig. 17. Sample collection of three different 3D objects: Top row: soccer ball; Middle row: cube; Bottom row: tetrahedron. The size, position and spatial orientation of the shapes, as well as the initial antennal location were randomized.

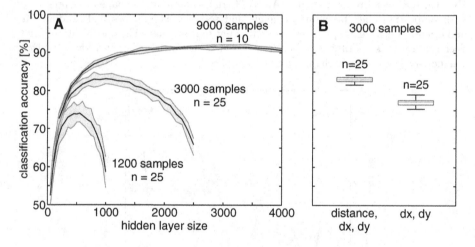

Fig. 18. 3D shape classification performance strongly depends on the sample size. **A**: Performance for three different sample sizes. A neural network requires a fairly large sample size to achieve high performance. It needs to "see" several contour projections of the randomly oriented objects. **B**: Influence of the input components on performance. Performance is still good with the velocity components of the Hopf oscillator, only.

low-frequency searching-behaviour to a low-amplitude, high-frequency sampling of an object. Although, in principle, it is possible to apply Contour-net to an array of tactile sensors, arbitrary contours may be sampled by a single tactile probe, for example an insect-like sensorised antenna as proposed by Patanè et al. (2012). With regard to tactile shape recognition, this may simplify the design and control of the sensing device, as it offers an alternative to the acquisition of a "tactile snapshot" by a sensor array. In the simplest case, Contour-net only uses a single binary value (contact or no-contact) and fixed phase-forwarding for contour tracing. Although this was shown to work for arbitrary object contours, the scan path, i.e. the spacing of the contacts, may become fairly irregular. This can be avoided by incorporating information about the surface normal, thus correcting the phase-forwarding according to the angle of attack measured during a contact. This can significantly improve the shape recognition performance of the subsequent machine learning algorithms. Moreover, it has been shown that perpendicular contacts of a tactile probe effectively reduces slip (Kaneko et al., 1995). There are different possibilities of determining the surface normal, e.g., by local re-sampling and/or inclusion of the current phase of the oscillator. As yet, a fast and reliable hardware method for detecting an object's surface normal is missing. Extracting the principal component orientation (see Patanè et al., 2012) from the two axis readings of the tip-mounted accelerometer might be a solution.

Contour-net yields a sequence of contact events that may be used for shape classification with minimal pre-processing (scaling, only). A comparison of three classifiers demonstrated that the information supplied by Contour-net was sufficient to achieve position-, size- and rotation-invariant shape recognition beyond 90 % for four alternative object shapes. Among the three approaches tested, SVM was found to achieve the most robust classification results with smaller training data sets. Using a SVM classifier, we could show that the shape classification works well even when exposed to unknown, morphed shapes. In general, it would be interesting to compare the system's classification performance with that of human subjects.

Advanced classification algorithms with sophisticated pre-processing were shown to gain classification rates of over 99.6 % (Ranzato et al., 2006; Ciresan et al., 2012) on a standard benchmark (MNISTdatabase, LeCun et al., 1998). Yet, plain, but big feed-forward networks with large hidden layers can achieve excellent classification results using standard online BP (Ciresan et al., 2010), but would require extremely long training times of several hours or days on a standard, year 2012 personal computer. A fast graphics card GPU implementation of the classic BP algorithm was presented in (Ciresan et al. 2010) to achieve acceptable training times (2 hours for an error below 1 %).

From the two ANN classifiers used in the present study, the ANN trained with the No-Prop-*fast* method does not reach quite as high classification rates. Yet, using the Matlab computing environment, it is easy to implement and substantially faster on the tactile shape recognition task (50 times faster than SVM_{rbf} and 6 times faster than BP_{scg}). Because of these properties, No-Prop-*fast* is a helpful tool for exploring and tuning parameters of the training data set at interactive speeds on standard PC hardware. Note that for large-scale data sets like ImageNet (Deng et al., 2009), the observations described above may no longer hold. High-performance implementations, using supercomputer clusters or multi-GPU computing (Ciresan et al., 2010; Krizhevsky et al., 2012), may scale better with the problem size.

Another possibility for further improvement of shape classification could be to extract higher-level features from collected contact events, and to use such pre-processed information instead of the scaled raw data used here. We expect that this should significantly reduce the required amount of training samples. Potentially suitable features might be the average Euclidean distance of contact events and the spread of Hopf oscillator phase differences between successive contact events.

In summary, the simulation experiments have shown that a position-, orientation- and size-invariant classification of 3D objects can be obtained even with incomplete scans of an object (see Sect. 3.4). The advantage of using an active beam-like tactile sensor is the lower complexity and cost, compared to whisker array approaches (Pearson et al., 2013) that instead can provide a more instantaneous tactile snapshot of the object. Beam-like tactile sensors are not only interesting for object negotiating in mobile robotics, but also have potential applications in industry. For example, a straight, filiform sensor can detect surface irregularities at drilled holes (burrs) with a high spatial resolution below $5\,\mu m$ (Kaneko and Tsuji, 2000). Another sensor inspired by a rat's vibrissa can sense geometrical parameters of deep microholes (location and taper of the side walls, Ju and Ling, 2015).

The robustness and simplicity of Contour-net will enable its implementation on a robotic platform such as the mobile robot (Erratic, Videre Design LLC) presented in Hoinville et al. (2014). This robot was equipped with two robotic feelers, that carry acceleration sensors at their tip, enabling not only a binary contact detection, but also a vibration frequency based estimation of the contact distance calculated offline (Patanè et al., 2012). Given the reasonably accurate localization of contour points in 3D space (Hoinville et al., 2014), we are confident that Contour-net and the subsequent incremental shape classification shown above can be transferred to robotic tactile sensing platforms with little requirement for platform-specific adaptation.

Acknowledgements. This work was supported by EU grant EMICAB (FP7-ICT, grant no. 270182) to Prof. Volker Dürr. We thank Thierry Hoinville for many important suggestion regarding the model, and Holk Cruse for valuable comments on earlier versions of the manuscript.

Appendix

The following link provides a zip archive that contains the Matlab scripts and data sets that were used to train the neural networks and support vector machines. Further, scripts are included to generate custom training data sets using the Contour-net method. http://www.andre-krause.net/publications/tcci15krause.zip

References

Bellman, R.E.: Adaptive Control Processes: A Guided Tour, vol. 4. Princeton University Press, Princeton (1961)

Brambilla, G., Buchli, J., Ijspeert, A.J.: Adaptive four legged locomotion control based on nonlinear dynamical systems. In: Nolfi, S., Baldassarre, G., Calabretta, R., Hallam, J.C.T., Marocco, D., Meyer, J.-A., Miglino, O., Parisi, D. (eds.) SAB 2006. LNCS (LNAI), vol. 4095, pp. 138–149. Springer, Heidelberg (2006)

Buchli, J., Righetti, L., Ijspeert, A.J.: A dynamical systems approach to learning: a frequency-adaptive hopper robot. In: Capcarrère, M.S., Freitas, A.A., Bentley, P.J., Johnson, C.G., Timmis, J. (eds.) ECAL 2005. LNCS (LNAI), vol. 3630, pp. 210–220. Springer, Heidelberg (2005)

Ciresan, D., Meier, U., Schmidhuber, J.: Multi-column deep neural networks for image classification. In: 2012 IEEE Conference on Computer Vision and Pattern Recognition (CVPR), pp. 3642–3649. IEEE (2012)

Ciresan, D.C., Meier, U., Gambardella, L.M., Schmidhuber, J.: Deep, big, simple neural nets for handwritten digit recognition. Neural Comput. **22**, 3207–3220 (2010)

Cruse, H., Kindermann, T., Schumm, M., Dean, J., Schmitz, J.: Walknet - a biologically inspired network to control six-legged walking. Neural Netw. **11**, 1435–1447 (1998)

Deng, J., Dong, W., Socher, R., Li, L.-J., Li, K., Fei-Fei, L.: Imagenet: a large-scale hierarchical image database. In: 2009 IEEE Conference on Computer Vision and Pattern Recognition, CVPR 2009, pp. 248–255. IEEE (2009)

Dürr, V., Ebeling, W.: The behavioural transition from straight to curve walking: kinetics of leg movement parameters and the initiation of turning. J. Exp. Biol. **208**, 2237–2252 (2005). doi:10.1242/jeb.01637

Dürr, V., König, Y., Kittmann, R.: The antennal motor system of the stick insect Carausius morosus: Anatomy and antennal movement pattern during walking. J. Comp. Physiol. A **187**(2), 131–144 (2001). doi:10.1007/s003590100183

Dürr, V., Schmitz, J., Cruse, H.: Behaviour-based modelling of hexapod locomotion: linking biology and technical application. Arthropod Struct. Dev. **33**(3), 237–250 (2004). doi:10.1016/j.asd.2004.05.004

Erber, J., Kierzek, S., Sander, E., Grandy, K.: Tactile learning in the honeybee. J. Comp. Physiol. A **183**, 737–744 (1998)

Esslen, J., Kaissling, K.-E.: Zahl und verteilung antennaler sensillen bei der honigbiene (Apis mellifera l.). Zoomorphology **83**, 227–251 (1976). doi:10.1007/BF00993511

Harischandra, N., Krause, A.F., Dürr, V.: Stable phase-shift despite quasi-rhythmic movements: a CPG-driven dynamic model ofactive tactile exploration in an insect. Front. Comput. Neurosci. **9**(107) (2015). doi:10.3389/fncom.2015.00107

Hoinville, T., Harischandra, N., Krause, A.F., Dürr, V.: Insect-inspired tactile contour sampling using vibration-based robotic antennae. In: Duff, A., Lepora, N.F., Mura, A., Prescott, T.J., Verschure, P.F.M.J. (eds.) Living Machines 2014. LNCS, vol. 8608, pp. 118–129. Springer, Heidelberg (2014). http://dx.doi.org/10.1007/978-3-319-09435-9_11

Hoinville, T., Wehner, R., Cruse, H.: Learning and retrieval of memory elements in a navigation task. In: Prescott, T.J., Lepora, N.F., Mura, A., Verschure, P.F.M.J. (eds.) Living Machines 2012. LNCS, vol. 7375, pp. 120–131. Springer, Heidelberg (2012)

Huang, G.-B.: An insight into extreme learning machines: random neurons, random features and kernels. Cogn. Comput. **6**(3), 376–390 (2014)

Huang, G.-B., Zhou, H., Ding, X., Zhang, R.: Extreme learning machine for regression and multiclass classification. IEEE Trans. Syst. Man Cybern. Part B: Cybern. **42**(2), 513–529 (2012)

Huang, G.-B., Zhu, Q.-Y., Siew, C.-K.: Extreme learning machine: theory and applications. Neurocomputing **70**(1–3), 489–501 (2006)

Ijspeert, A.J.: Central pattern generators for locomotion control in animals and robots: a review. Neural Netw. **21**(4), 642–653 (2008)

Ijspeert, A.J., Crespi, A., Ryczko, D., Cabelguen, J.-M.: From swimming to walking with a salamander robot driven by a spinal cord model. Science **315**(5817), 1416–1420 (2007)

Ju, F., Ling, S.-F.: Bioinspired active whisker sensor for geometry detection of high aspect ratio microholes with simultaneous actuation and sensing capability. Smart Mater. Struct. **24**(3) (2015)

Kaneko, M., Kanayama, N., Tsuji, T.: 3-D active antenna for contact sensing. IEEE Int. Conf. Robot. Autom. (ICRA) **1**, 1113–1119 (1995)

Kaneko, M., Kanayma, N., Tsuji, T.: Active antenna for contact sensing. IEEE Trans. Robot. Autom. **14**(2), 278–291 (1998)

Kaneko, M., Tsuji, T.: A whisker tracing sensor with 5 μm sensitivity. IEEE Int. Conf. Robot. Autom. (ICRA) **4**, 3907–3912 (2000)

Kevan, P.G., Lane, M.A.: Flower petal microtexture is a tactile cue for bees. Proc. Nat. Acad. Sci. **82**(14), 4750–4752 (1985)

Kim, D., Möller, R.: Biomimetic whiskers for shape recognition. Robot. Autonom. Syst. **55**(3), 229–243 (2007)

Kohavi, R.: A study of cross-validation and bootstrap for accuracy estimation and model selection. Proc. Int. Joint Conf. Artif. Intell. (IJCAI) **14**, 1137–1145 (1995)

Krause, A.F., Dürr, V.: Tactile efficiency of insect antennae with two hinge joints. Biol. Cybern. **91**(3), 168–181 (2004). doi:10.1007/s00422-004-0490-6

Krause, A.F., Dürr, V.: Active tactile sampling by an insect in a step-climbing paradigm. Front. Behav. Neurosci. **6**(30), 1–17 (2012). doi:10.3389/fnbeh.2012.00030

Krause, A.F., Essig, K., Essig-Shih, L.-Y., Schack, T.: Classifying the differences in gaze patterns of alphabetic and logographic L1 readers – a neural network approach. In: Iliadis, L., Jayne, C. (eds.) EANN/AIAI 2011, Part I. IFIP AICT, vol. 363, pp. 78–83. Springer, Heidelberg (2011)

Krause, A.F., Essig, K., Piefke, M., Schack, T.: No-prop-fast - a high-speed multilayer neural network learning algorithm: mnist benchmark and eye-tracking data classification. In: Iliadis, L., Papadopoulos, H., Jayne, C. (eds.) Engineering Applications of Neural Networks (EANN 2013), pp. 446–455. Springer, Berlin Heidelberg (2013)

Krause, A.F., Winkler, A., Dürr, V.: Central drive and proprioceptive control of antennal movements in the walking stick insect. J. Physiol. Paris **107**(1–2), 116–129 (2013b). doi:10.1016/j.jphysparis.2012.06.001

Krizhevsky, A., Sutskever, I., Hinton, G.E.: Imagenet classification with deep convolutional neural networks. In: Bartlett, P. (ed.) Advances in Neural Information Processing Systems 25: 26th Annual Conference on Neural Information Processing Systems 2012, pp. 1097–1105. Curran Associates Inc, Red Hook (2012)

LeCun, Y., Bottou, L., Bengio, Y., Haffner, P.: Gradient-based learning applied to document recognition. Proc. IEEE **86**(11), 2278–2324 (1998)

Lederman, S., Klatzky, R.: Haptic perception: a tutorial. Attention Percept. Psychophys. **71**(7), 1439–1459 (2009). doi:10.3758/APP.71.7.1439

Legland, D.: Geom2d toolbox (2012). http://matgeom.sourceforge.net/

Møller, M.F.: A scaled conjugate gradient algorithm for fast supervised learning. Neural Netw. **6**(4), 525–533 (1993)

Patanè, L., Hellbach, S., Krause, A.F., Arena, P., Dürr, V.: An insect-inspired bionic sensor for tactile localization and material classification with state-dependent modulation. Front. Neurorobot. **6**(8), 1–18 (2012). doi:10.3389/fnbot.2012.00008

Pearson, M.J., Fox, C., Sullivan, J.C., Prescott, T.J., Pipe, T., Mitchinson, B.: Simultaneous localisation and mapping on a multi-degree of freedom biomimetic whiskered robot. In: Antonelli, G., et al. (eds.) 2013 IEEE International Conference on Robotics and Automation (ICRA), pp. 586–592. IEEE (2013)

Pearson, M.J., Pipe, A.G., Melhuish, C., Mitchinson, B., Prescott, T.J.: Whiskerbot: a robotic active touch system modeled on the rat whisker sensory system. Adapt. Behav. **15**(3), 223–240 (2007)

Prescott, T.J., Dürr, V.: The world of touch. Scholarpedia **10**(4), 32688 (2015). doi:10.4249/scholarpedia.32688

Ranzato, M., Poultney, C., Chopra, S., LeCun, Y.: Efficient learning of sparse representations with an energy-based model. In: Platt, J. (ed.) Advances in Neural Information Processing Systems (NIPS 2006), vol. 19. MIT Press, Cambridge (2006)

Refaeilzadeh, P., Tang, L., Liu, H.: Cross-validation. Encyclopedia of Database Systems, pp. 532–538. Springer, Berlin (2009)

Righetti, L., Ijspeert, A.J.: Programmable central pattern generators: an application to biped locomotion control. In: Hutchinson, S., et al. (eds.) IEEE International Conference on Robotics and Automation (ICRA), pp. 1585–1590. IEEE (2006)

Russell, R.A., Wijaya, J.A.: Object location and recognition using whisker sensors. In: Roberts, J., Wyeth, G. (eds.) Australasian Conference on Robotics and Automation, pp. 761–768 (2003)

Schilling, M., Hoinville, T., Schmitz, J., Cruse, H.: Walknet, a bio-inspired controller for hexapod walking. Biol. Cybern. **107**(4), 397–419 (2013). doi:10.1007/s00422-013-0563-5

Schütz, C., Dürr, V.: Active tactile exploration for adaptive locomotion in the stick insect. Philos. Trans. R. Soc. B: Biol. Sci. **366**(1581), 2996–3005 (2011). doi:10.1098/rstb.2011.0126

Solomon, J.H., Hartmann, M.J.: Biomechanics: robotic whiskers used to sense features. Nature **443**(7111), 525 (2006)

Staudacher, E., Gebhardt, M.J., Dürr, V.: Antennal movements and mechanoreception: neurobiology of active tactile sensors. Adv. Insect Physiol. **32**, 49–205 (2005). doi:10.1016/S0065-2806(05)32002-9

Sullivan, J., Mitchinson, B., Pearson, M.J., Evans, M., Lepora, N.F., Fox, C.W., Melhuish, C., Prescott, T.J.: Tactile discrimination using active whisker sensors. IEEE Sens. J. **12**(2), 350–362 (2012)

Widrow, B., Greenblatt, A., Kim, Y., Park, D.: The No-Prop algorithm: a new learning algorithm for multilayer neural networks. Neural Netw. **37**, 182–188 (2013)

Real-Time Tear Film Classification Through Cost-Based Feature Selection

Verónica Bolón-Canedo$^{(\boxtimes)}$, Beatriz Remeseiro, Noelia Sánchez-Maroño, and Amparo Alonso-Betanzos

Departamento de Computación, Universidade da Coruña, A Coruña, Spain
{vbolon,bremeseiro,nsanchez,ciamparo}@udc.es

Abstract. Dry eye syndrome is an important public health problem, and can be briefly defined as a symptomatic disease which affects a wide range of population and has a negative impact on their daily activities. In clinical practice, it can be diagnosed by the observation of the tear film lipid layer patterns, and their classification into one of the Guillon categories. However, the time required to extract some features from tear film images prevents the automatic systems to work in real time. In this paper we apply a framework for cost-based feature selection to reduce this high computational time, with the particularity that it takes the cost into account when deciding which features to select. Specifically, three representative filter methods are chosen for the experiments: Correlation-Based Feature Selection (CFS), minimum-Redundancy-Maximum-Relevance (mRMR) and ReliefF. Results with a Support Vector Machine as a classifier showed that the approach is sound, since it allows to reduce considerably the computational time without significantly increasing the classification error.

Keywords: Tear film · Dry eye · Real time · Cost-based feature selection · Filter methods · Classification

1 Introduction

Tears are secreted from the lachrymal gland and distributed by blinking to form the tear film of the ocular surface [1]. The tear film covers the exposed anterior surface of the eye and is essential for the execution of its functions. Wolff provided the classical description of the preocular tear film as a three-layered structure which consists of an anterior lipid layer, an aqueous layer, and a deep mucin layer [2]. Each of them plays a different role towards the formation and stability of the structure.

Quantitative or qualitative changes in the normal lipid layer have a negative effect on the quality of vision, and on the evaporation of tears from the ocular surface [3]. In fact, a substantial tear evaporation caused by alterations of the lipid layer is characteristic of the *evaporative dry eye* (EDE). Note that dry eye is a prevalent disease which leads to irritation of the ocular surface, and is

© Springer-Verlag Berlin Heidelberg 2015
N.T. Nguyen et al. (Eds.): TCCI XX, LNCS 9420, pp. 78–98, 2015.
DOI: 10.1007/978-3-319-27543-7_4

associated with symptoms of discomfort and dryness [4]. A fourth of the patients who visit ophthalmic clinics report symptoms of dry eye, making it one of the most common conditions seen by eye care practitioners [5]. Moreover, the current work conditions, such as computer use, have increased the proportion of people who suffer this disease [4].

The diagnosis of EDE is complicated since it has no single characteristic sign or symptom, and no single diagnostic measure. Therefore, several tests are necessary in order to obtain a clear diagnosis. To evaluate different aspects of the tear film, there are a wide number of tests, which can be divided into two main groups: quantitative tear film tests, which assess tear secretion; and qualitative tear film tests, which measure the ability of the tear film to remain stable. The lipid layer assessment is one of the most popular qualitative tests, and it assesses tear film quality and the lipid layer thickness by non-invasively imaging the superficial lipid layer with interferometry. The Tearscope Plus [6] is the instrument of choice for rapid assessment of lipid layer thickness, and allows the qualitative analysis of the lipid layer structure. This test is based on a standard classification defined by Guillon [7], who specified various types of lipid layer patterns. There is no doubt that this medical proof is a valuable test which provides noteworthy information by using non-invasive procedures. However, many eye care practitioners have abandoned this test because the difficulty in interpreting the lipid layer patterns, specially the thinner ones. This is why automatic tear film lipid layer classification could become a key step to diagnose EDE, since it eliminates the subjectivity of the manual process and saves time for experts.

First attempts to automatize tear film classification can be found in [8], in which it was demonstrated how the interference phenomena can be characterized as a color texture pattern. These results were later improved by using a wide set of texture analysis techniques and color spaces, and extended to different machine learning algorithms [9,10]. The problem with these previous approaches, which prevented their clinical use, is that they required a large amount of time for the computation of the features.

The presence of multiple patterns in the tear film lipid layer, observed in some patients, may affect the classification of a patient's image into a single Guillon category, as the previous approaches do. Alternatively, performing local analysis of the images to detect multiple categories per patient may be more accurate in some cases, and it would be useful to discern different local states. For this reason, the so-called tear film maps were presented in [11,12] to illustrate the distribution of different patterns over the tear film. In computational terms, the creation of these maps involves a high increase in the memory and time requirements, since it is based on the application of the previous approaches but at a local level by means of different image segmentation techniques. Consequently, the problem of extracting features in the lowest time is of crucial importance.

In order to deal with this problem, feature selection can play a crucial role. In machine learning, *feature selection* (FS) is defined as the process of detecting the relevant features and discarding the irrelevant ones [13], to obtain the subset of features that describes properly the given problem. Because the number of features to extract and process will be reduced, the time required will be also reduced in consonance, and most frequently, this can be achieved with a minimum degradation of performance.

The most common approaches in feature selection are to find either a subset of features that maximizes a given metric, or either an ordered ranking of the features according to some merit. In this manner, several feature selection methods were applied to the problem of tear film classification in [14]. However, we are not only interested in maximizing the merit of a subset of features, but also in reducing the costs that may be associated to features (in this case, in the form of computational time). Consequently, the above mentioned research proposed an ad-hoc method based on Correlation-based Feature Selection (CFS) which manually rejected some of the most expensive features.

Trying to solve this problem, this paper deals with cost-based feature selection by using a general framework proposed by the authors in [15] with the aim of obtaining an adequate subset of features in a short time. Methods in this framework can be employed to achieve a trade-off between the feature selection metric and the cost associated to the selected features, in order to select relevant features with a low associated cost. In a previous work, the adequacy of this type of feature selection with the implementation of ReliefF [16] was demonstrated in the problem at hand. In this work, we also apply other two state-of-the-art methods, Correlation-based Feature Selection and minimum Redundancy Maximum Relevance, trying to improve the experimental results.

This paper is organized as follows: Sect. 2 describes the problem of tear film lipid layer classification, Sect. 3 introduces the concept of cost-based feature selection, and Sects. 4 and 5 present the experimental setup and experimental results, respectively. Finally, Sect. 6 includes the conclusions.

2   Tear Film Lipid Layer Classification

Optometrists carry out tear film assessment by means of the evaluation of the lipid layer through a manual process, which consists in classifying images obtained with the Tearscope Plus into one of the Guillon categories (see Table 1). The Tearscope Plus has proven its validity to the lipid layer pattern assessment [17,18]. As stated before, the test has declined its use by the clinical eye care professionals due to two main reasons: (1) the lipid layer patterns, specially the thinner ones, are very difficult to interpret, and (2) a huge bank of images for reference purposes is lacking. Nevertheless, there is no doubt that the examination of the structure of the tear film lipid layer is a valuable technique which provides practitioners with relevant information about the stability of the tear film by using non-invasive procedures.

This clinical task is not only difficult and time-consuming, but also affected by the subjective interpretation of the observers. This has motivated the development of automated techniques to characterize the interference phenomena characteristic of the lipid layer patterns, in such a way that the tear film lipid

Table 1. Appearance and estimated thickness of the tear film lipid layer patterns observed with the Tearscope Plus

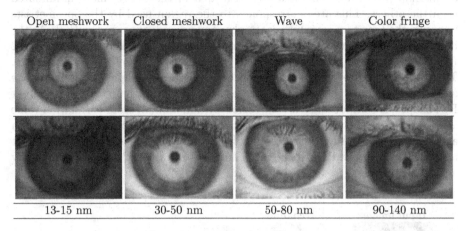

Open meshwork	Closed meshwork	Wave	Color fringe
13-15 nm	30-50 nm	50-80 nm	90-140 nm

layer can be automatically classified into one of the categories enumerated by Guillon. In this sense, this section presents a methodology which, from a photography of the eye, detects a region of interest (ROI) and extracts its low-level features, generating a feature vector which describes it, to be finally classified into one of the target categories. Figure 1 illustrates these three steps, which will be subsequently presented in depth.

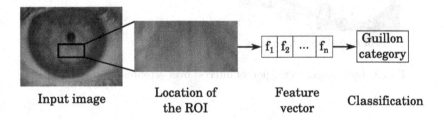

Input image   Location of the ROI   Feature vector   Classification

Fig. 1. Steps of the research methodology for tear film classification

2.1   Location of the Region of Interest

The input images, as depicted in Fig. 2, include several areas of the eye which do not contain relevant information for the classification, such as the sclera, eyelids and eyelashes. Optometrists who analyze these images usually focus their attention on the bottom part of the iris, since this is the area in which the tear can be perceived with the highest contrast. This forces a preprocessing step aimed at extracting the region in which the tear film classification takes place, known as *region of interest* (ROI) [10].

Figure 2 depicts an example of the process performed to locate the ROI. The acquisition procedure generates a central area in the image, more illuminated than the others. This area corresponds to the region used by the optometrists to assess the tear film by interference phenomena and, thus, to the ROI. As the illumination plays an essential role, the input image in RGB is transformed to the Lab color space and only its component of luminance L is selected in this stage. Then, the normalized cross-correlation between the L component of the image and a set of ring-shaped templates previously generated, that cover the different ROI shapes, is computed. Next, the region with maximum cross-correlation value is selected and, as the region of interest is situated at the bottom part, the top area is rejected. Finally, the rectangle of maximum area inside this bottom part is located and so the ROI of the input image is obtained through a completely automatic process.

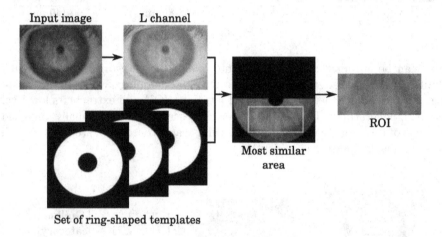

Fig. 2. Location of the region of interest over a representative image

2.2 Feature Vector

After extracting the ROI, the next step entails analyzing its low-level features by means of color and texture analysis, since both color and interference patterns are the two discriminant features of the Guillon categories for tear film classification.

Color Analysis: Lab Color Space. Color is one of the discriminant features of the Guillon categories for tear film classification. Some categories show distinctive color characteristics and, for this reason, tear film images were analyzed [10] not only in grayscale but also in the Lab and RGB color spaces. In this previous work, the best results were obtained in Lab, so we will focus on it from now on.

The CIE 1976 L*a*b* color space [19] (Lab) is a chromatic color space which describes all the colors that the human eye can perceive. It was defined by the International Commission on Illumination, abbreviated as CIE from its French title *Commission Internationale de l'Eclairage*. Lab is a 3D model where its three coordinates represent: the luminance of the color L, its position between magenta and green a, and its position between yellow and blue b. Its use is recommended by CIE in images with natural illumination. In addition, this color space is perceptually uniform, which means that a change of a certain amount in a color value produces a change of the same visual relevance. This characteristic is also important since the specialists' perception is being imitated.

Texture Analysis: Co-Occurrence Features. As well as color, interference patterns are one of the discriminant features of the Guillon categories. In [10], it has been demonstrated that the interference patterns can be characterized by means of texture features, since thick lipid layers show clear patterns while thinner layers are more homogeneous. Additionally, previous results shown that the co-occurrence features technique provides the most competitive results for this problem, thus we will consider it in this stage.

Co-occurrence features analysis [20] is a popular and effective texture descriptor based on the computation of the conditional joint probabilities of all pairwise combinations of gray levels, given an interpixel distance d and an orientation θ. The method generates a set of *gray level co-occurrence matrices* (GLCM), and extracts several statistical measures from their elements. Using the *Chebyshev* distance, the number of orientations and, accordingly, the number of matrices for a distance d is $4d$.

From each GLCM, a set of 14 statistics proposed by Haralick et al. [20] are computed. For explanatory purposes, the definition of 2 of these 14 statistical measures is shown:

$$f_1 = \sum_{i=1}^{N} \sum_{j=1}^{N} \left(\frac{P_{\theta,d}(i,j)}{R} \right)^2 \tag{1}$$

$$f_2 = \sum_{n=0}^{N-1} n^2 \left\{ \sum_{|i-j|=n} \left(\frac{P_{\theta,d}(i,j)}{R} \right) \right\} \tag{2}$$

where $P_{\theta,d}(i,j)$ are the elements of the GLCM, N is the number of distinct gray levels in the input image, and R is a normalizing constant. The angular second-moment feature f_1 is a measure of homogeneity of the image, and the contrast feature f_2 is a measure of the amount of local variations present in the image. Finally, the mean and the range of these 14 statistics are calculated across matrices and a set of 28 features composes the texture descriptor for a distance d.

Definition of the Feature Vector. The feature vector of an input image is created by using the color model and texture descriptor previously presented.

The use of the Lab color space entails converting the three channels of the RGB image into the three components of Lab [21]. Then, each component is analyzed separately and its texture descriptor is obtained. Therefore, the final descriptor is the concatenation of the three descriptors. Note that when different distances are used, their individual descriptors are combined by means of their concatenation.

2.3   Classification: Support Vector Machine

Supervised learning entails learning a mapping between a set of input features and output labels, and applying this mapping to predict the outputs for new data [22]. The resulting classifier is then used to assign class labels to the new instances where the values of the features are known, but the value of the class label is unknown. Several popular machine learning algorithms were applied to tear film classification in [9], and the support vector machine demonstrated to be the most competitive classifier for the problem at hand.

Support vector machine (SVM) [23] is a widely-used classifier based on the statistical learning theory and revolves around the notion of a "margin", either side of a hyperplane that separates two classes. Most real world problems involve non-separable data for which no hyperplane exists that successfully separates two classes. In this case, the idea is to map the input data onto a higher dimensional space and define a separating hyperplane there. The higher-dimensional space is called the transformed feature space and it is obtained using *kernel functions*.

SVM necessarily reaches a global minimum and avoids ending in a local minima, which may happen in other algorithms. They avoid problems of overfitting and, with an appropriate kernel, they can work well even if the data is not linearly separable.

3   Cost-Based Feature Selection

Feature selection methods can be divided into wrappers, filters and embedded methods [13]. The filter model relies on the general characteristics of training data and carries out the feature selection process as a pre-processing step with independence of the induction algorithm. The embedded methods generally perform feature selection in the process of training and are specific to given learning machines. Wrappers, in turn, involve optimizing a predictor as part of the selection process. Wrappers and embedded methods tend to obtain better performances but at the expense of being very time consuming and having the risk of overfitting when the sample size is small. In contrast, filters are faster, easier to implement, scale up better than wrappers and embedded methods, and can be used as a pre-processing step before applying other more complex methods.

New feature selection methods are continuously appearing covering the three approaches aforementioned, however, the great majority of them only focus on removing irrelevant and redundant features but not on the costs for obtaining the input features. The cost associated to a feature can be related to different concepts. For example, in medical diagnosis, a pattern consists of observable symptoms (such as age, sex, etc.) along with the results of some diagnostic tests. Contrary to observable symptoms, which have no cost, diagnostic tests have associated costs and risks. For example, an invasive exploratory surgery is much more expensive and risky than a blood test [24]. Also, the risk of extracting a feature can be found in the work presented by Bahamonde et al. [25], in which for evaluating the merits of beef cattle as meat producers is necessary to carry out zoometry on living animals.

However, the cost can also be related to computational issues. In the medical imaging field, extracting a feature from a medical image can have a high computational cost. For example, in the case at hand, the computational cost for extracting each feature required by co-occurrence features [26] is not the same, which implies different computational times. In other cases, such as real-time applications, the space complexity is negligible, but the time complexity is very important [27].

As one may notice, features with an associated cost can be found in many real-life applications. However, this has not been the focus of much attention for feature selection researchers. Despite the existence of previous attempts in classification and feature extraction [28–31], to the best knowledge of the authors, there are only a few attempts to deal with this issue in feature selection. In the early 90 s, Feddema et al. [27] were developing methodologies for the automatic selection of image features to be used by a robot. For this selection process, they employed a weighted criterion that took into account the computational expense of features, i.e., the time and space complexities of the feature extraction process. Several years later, Yang et al. [24] proposed a genetic algorithm to perform feature subset selection where the fitness function combined two criteria: the accuracy of the classification function realized by the neural network and the cost of performing the classification (defined by the cost of measuring the value of a particular feature needed for classification, the risk involved, etc.). A similar approach was presented by Huang et al. [32], in which a genetic algorithm is used for feature selection and parameters optimization for a support vector machine. In this case, classification accuracy, the number of selected features and the feature cost were the three criteria used to design the fitness function. In [33], a hybrid method for feature subset selection based on ant colony optimization and artificial neural networks was presented. The heuristic that enables ants to select features is the inverse of the cost parameter.

The methods found in the literature that deal with cost associated to the features, which were described above, have the disadvantage of being computationally expensive by having interaction with a classifier, which prevents their use in large databases, a trending topic in recent years [34]. However, in [15] we proposed a framework for feature selection applied together with the filter model, which is known to have a low computational cost and be independent of any classifier. By being fast and with a good generalization ability, filters using this cost-based feature selection framework are suitable for application to databases with a great number of input features, and its adequacy was demonstrated in scenarios such as microarray DNA data [15]. Later, this framework modified for applying the well-known ReliefF filter was tested on the problem of tear film lipid layer classification, obtaining promising results [16].

3.1   Cost-Based Methods Employed

In [15], we have presented a framework for cost-based feature selection, which could be applicable to any filter. To be applied to the problem of tear film lipid layer classification, we have chosen three representative filters to implement the framework and carry out the experimentation. The filters chosen are CFS, which is a subset filter, and mRMR and ReliefF, which are ranker filters.

Minimum Cost CFS, mC-CFS. Correlation-based feature selection (CFS) [35] is a multivariate subset filter algorithm. It uses a search algorithm combined with an evaluation function to estimate the merit of feature subsets. The implementation of CFS utilized in this work uses forward best first search [36] as its search algorithm. Best first search is an artificial intelligence search strategy that allows backtracking along the search path. It moves through the search space by making local changes to the current feature subset. If the explored path looks uninteresting, the algorithm can backtrack to a previous subset and continue the search from there on. As a stopping criterion, the search terminates if five consecutive fully expanded (all possible local changes considered) subsets show no improvement over the current best subset.

The evaluation function takes into account the usefulness of individual features for predicting the class label as well as the level of correlation among them. It is assumed that good feature subsets contain features highly correlated with the class and uncorrelated with each other. The evaluation function can be seen in (3).

$$M_S = \frac{k \overline{r_{ci}}}{\sqrt{k + k(k-1)\overline{r_{ii}}}} \tag{3}$$

where M_S is the merit of a feature subset S that contains k features, $\overline{r_{ci}}$ is the average correlation between the features of S and the class, and $\overline{r_{ii}}$ is the average intercorrelation between the features of S. In fact, this function is Pearson's correlation with all variables standardized. The numerator estimates how S can predict the class and the denominator quantifies the redundancy among the features in S.

The modification of CFS consists of adding a term to the evaluation function to take into account the cost of the features, as can be seen in Eq. (4).

$$MC_S = \frac{k \overline{r_{ci}}}{\sqrt{k + k(k-1)\overline{r_{ii}}}} - \lambda \frac{\sum_{i=1}^{k} C_i}{k} \tag{4}$$

where MC_S is the merit of the subset S affected by the cost of the features, C_i is the cost of the feature i, and λ is a parameter introduced to weight the influence of the cost in the evaluation function.

The parameter λ is a positive real number. If $\lambda = 0$, the cost is ignored and the method works as the regular CFS. If $0 < \lambda < 1$, the influence of the cost is smaller than the other term. If $\lambda = 1$ both terms have the same influence and if $\lambda > 1$, the influence of the cost is greater than the influence of the other term.

Minimum Cost mRMR, mC-mRMR. Minimum Redundancy Maximum Relevance (mRMR) [37] is a multivariate ranker filter algorithm. As mRMR is a ranker, the search algorithm is simpler than that of CFS. The evaluation function combines two constraints (as the name of the method indicates), maximal relevance and minimal redundancy. The former is denoted by the letter D, it corresponds with the mean value of all mutual information values between each feature x_i and class c, and has the expression shown in Eq. (5).

$$D(S, c) = \frac{1}{|S|} \sum_{x_i \in S} I(x_i; c) \tag{5}$$

where S is a subset of features and $I(x_i; c)$ is the mutual information between the feature x_i and the class c. The expression of $I(x; y)$ is shown in Eq. (6).

$$I(x; y) = \int \int p(x, y) \log \frac{p(x, y)}{p(x)p(y)} dx dy \tag{6}$$

The constraint of minimal redundancy is denoted by the letter R, and has the expression shown in (7).

$$R(S) = \frac{1}{|S|^2} \sum_{x_i, x_j \in S} I(x_i, x_j) \tag{7}$$

The evaluation function to be maximized combines the two constraints (5) and (7). It is called *minimal-redundancy-maximal-relevance* (mRMR) and has the expression shown in (8).

$$\Phi(D, R) = \frac{1}{|S|} \sum_{x_i \in S} I(x_i; c) - \frac{1}{|S|^2} \sum_{x_i, x_j \in S} I(x_i, x_j) = D(S, c) - R(S) \tag{8}$$

In practice, it is an incremental search method that selects on each iteration the feature that maximizes the evaluation function. Assume we already have S_{m-1}, the feature set with $m - 1$ features, the mth selected feature will optimize the following condition:

$$\max_{x_j \in X - S_{m-1}} \left[I(x_j; c) - \frac{1}{m-1} \sum_{x_i \in S_{m-1}} I(x_j; x_i) \right] \tag{9}$$

The modification of mRMR which is proposed consists of adding a term to the condition to be maximized so as to take into account the cost of the feature to be selected, as can be seen in (10).

$$\max_{x_j \in X - S_{m-1}} \left[\left(I(x_j; c) - \frac{1}{m-1} \sum_{x_i \in S_{m-1}} I(x_j; x_i) \right) - \lambda C_j \right] \qquad (10)$$

where C_j is the cost of the feature j, and λ is a parameter introduced to weight the influence of the cost in the evaluation function, as explained in the previous subsection.

Minimum Cost ReliefF, mC-ReliefF. ReliefF [38] is a multivariate ranker filter algorithm derived from the original version Relief [39]. It randomly selects an instance R_i, but then searches for k of its nearest neighbors from the same class c, nearest hits H_j, and also k nearest neighbors from each one of the different classes, nearest misses $M_j(c)$. It updates the quality estimation $W[A]$ for all attributes A depending on their values for R_i, hits H_j and misses $M_j(c)$. If instances R_i and H_j have different values of the attribute A, then this attribute separates instances of the same class, which clearly is not desirable, and thus the quality estimation $W[A]$ has to be decreased. On the contrary, if instances R_i and M_j have different values of the attribute A for a class then the attribute A separates two instances with different class values which is desirable so the quality estimation $W[A]$ is increased. Since ReliefF considers multiclass problems, the contribution of all the hits and all the misses is averaged. Besides, the contribution for each class of the misses is weighted with the prior probability of that class $P(c)$ (estimated from the training set). The whole process is repeated m times (where m is a user-defined parameter) and can be seen in Algorithm 1.

Algorithm 1. Pseudo-code of ReliefF algorithm

Data: training set D, iterations m, attributes a
Result: the vector W of estimations of the qualities of attributes

1 set all weights $W[A] := 0$
2 **for** $i \leftarrow 1$ **to** m **do**
3   randomly select an instance R_i
4   find k nearest hits H_j
5   **for** *each class* $C \neq class(R_i)$ **do**
6 | from class C find k nearest misses $M_j(C)$
     end
end
7 **for** $f \leftarrow 1$ **to** a **do**
8       $W[f] :=$

     $W[f] - \frac{\sum_{j=1}^{k} diff(f, R_i, H_j)}{(m \cdot k)} + \frac{\sum_{C \neq class(R_i)} \left[\frac{P(C)}{1 - P(class(R_i))} \sum_{j=1}^{k} diff(f, R_i, M_j(C)) \right]}{(m \cdot k)}$

end

The function $diff(A, I_1, I_2)$ calculates the difference between the values of the attribute A for two instances, I_1 and I_2. If the attributes are nominal, it is defined as:

$$diff(A, I_1, I_2) = \begin{cases} 0; & value(A, I_1) = value(A, I_2) \\ 1; & otherwise \end{cases}$$

The modification of ReliefF with our cost-based framework, mC-ReliefF, consists of adding a term to the quality estimation $W[f]$ to take into account the cost of the features, as can be seen in (11).

$$W[f] := W[f] - \frac{\sum_{j=1}^{k} diff(f, R_i, H_j)}{(m \cdot k)} +$$
$$\frac{\sum_{c \neq class(R_i)} \left[\frac{P(c)}{1 - P(class(R_i))} \sum_{j=1}^{k} diff(f, R_i, M_j(c)) \right]}{(m \cdot k)} - \lambda \cdot C_f \tag{11}$$

where C_f is the cost of the feature f, and λ is a free parameter introduced to weight the influence of the cost in the quality estimation of the attributes. When $\lambda > 0$, the greater the λ the greater the influence of the cost.

4 Experimental Setup

The experiments consist of applying the proposed cost-based feature selection methods (mC-CFS, mC-mRMR and mC-ReliefF) to the real problem of tear film lipid layer classification using the dataset VOPTICAL_I1 [40]. It consists of 105 images (samples) belonging to the four Guillon's categories (classes). All these images have been annotated by optometrists from the Optometry Service of the University of Santiago de Compostela (Spain). The distribution of the four classes of the dataset can be seen in Table 2.

Table 2. Distribution of the four different classes in VOPTICAL_I1 dataset

Class	No. of samples
Open meshwork	29
Closed meshwork	29
Wave	25
Color fringe	22

The aim of the experiments is to study the behavior of the cost-based methods under the influence of the λ parameter, and to check if the time required to extract the features can be reduced without affecting significantly the error. Therefore, the performance is evaluated in terms of both the total cost of the selected features (in this case extracting time) and the classification error obtained by a Support Vector Machine (SVM).

The error and the cost will be estimated with a 10-fold cross validation. This technique consists of dividing the dataset into 10 subsets and repeating the process 10 times. Each time, 1 subset is used as the test set and the other 9 subsets are put together to form the training set. Finally, the average error and cost across all 10 trials are computed. It is expected that the larger the λ, the lower the cost and the higher the error, since increasing λ gives more weight to cost at the expense of reducing the importance of the relevance of the features.

Moreover, a Kruskal-Wallis statistical test and a multiple comparison test (based on Tukey's honestly significant difference criterion) [41] have been run on the errors and costs obtained. These results could help the user to choose the value of the λ parameter.

5   Experimental Results

This section presents the average cost and error for several values of λ and the three proposed cost-based feature selection methods. Since mC-mRMR and mC-ReliefF return an ordered ranking of features, a threshold is required. In this work, we have opted for retaining the top 25, 35 and 50 features in the obtained ranking.

Figure 3(a) shows the error/cost plot when applying mC-CFS with different values of λ. In the left axis, one can see the average test error on the 10 folds whereas, in the right axis, the average cost for the selected features is shown. Notice that, for the experiments, the different times associated to the features were normalized between 0 and 1.

As expected, the higher the λ, the lower the cost. However, the error is not so affected by the increase in the value of λ, which suggests that it is possible to select less expensive features without degrading the classification performance. Figure 4 displays the results of the Kruskal-Wallis statistical test with a level of

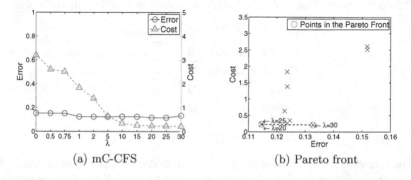

(a) mC-CFS   (b) Pareto front

Fig. 3. Error/cost plot (left) and Pareto front (right) for mC-CFS and different values of λ (Color figure online)

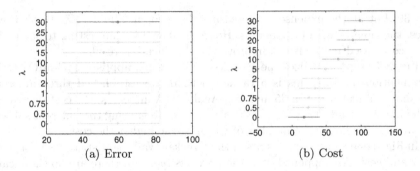

(a) Error   (b) Cost

Fig. 4. Statistical tests for mC-CFS and different values of λ

significance $\alpha = 0.05$ for both the error and the cost for the different values of λ tested. It can be seen that there are not significant differences between the different errors. On the contrary, the costs obtained when $\lambda \geq 10$ are significantly lower than that of the original version that does not take the cost into account ($\lambda = 0$). Thus, we can affirm that it is possible to decrease the cost without significantly increase the classification error.

Trying to shed light on the issue of which value of λ is the best for the problem at hand, the Pareto front [42] is shown in Fig. 3(b). In multi-objective optimization, the Pareto front is defined as the border between the region of feasible points, for which all constraints are satisfied, and the region of infeasible points. In this case, solutions are constrained to minimize classification error and cost. In Fig. 3(b), the points (values of λ) in the Pareto front are marked with a red circle. All those points get either the lowest cost or the lowest error, and it is decision of the users if they prefer to minimize either the cost or the classification error. In contrast, choosing a value of λ outside the Pareto front would imply choosing a worse solution than any in the Pareto front.

Table 3 reports the classification error and cost (in the form of the real time, i.e., not normalized between 0 and 1) for all the Pareto front points. As a 10-fold cross-validation was performed, the number of features and the time is the average on the 10 executions. As expected, the higher the λ, the higher the error and the lower the time. The best result in terms of classification error was obtained with $\lambda = 20$ and $\lambda = 30$. In turn, the lowest time was obtained with

Table 3. Mean classification error(%), time (milliseconds) and number of selected features on average for the 10 folds for the Pareto front points obtained by mC-CFS. Best error and time are marked in bold face.

λ	Error	Time	Features
20	**11.45**	400.59	27.40
25	**11.45**	390.93	26.90
30	13.27	**368.77**	25.40

$\lambda = 30$, but at the expense of increasing the error in almost 2 %. In all these cases, the time is already under 1 s. However, it will be interesting to see if the other cost-based methods can improve the classification results.

Figure 5 (top) visualizes the error/cost plots when applying mC-mRMR. As aforementioned, since this is a ranker method, we established three different thresholds of features (25, 35 and 50). Again, as λ increases, the cost decreases, without notably affecting the classification error. Regarding the different subsets of features, the larger the number of features, the higher the cost.

In Fig. 6 one can see the results of the Kruskal-Wallis statistical tests for both error and cost. As happened with the previous method, there are no significant differences between the errors obtained for different values of λ, for each number of selected features. Nevertheless, when λ is equal or greater to 5, the cost is significantly reduced.

Figure 5 (bottom) and Table 4 show the points in the Pareto front for each number of features to select. The best result in terms of error is obtained when $\lambda = 5$ and 50 features are retained. On the contrary, the lowest time –on average– is achieved when $\lambda = 30$ and 25 features are being selected, but at the expense of the worst classification error. Notice that, in this case, the times are in the order of tens of milliseconds, which makes this solution highly appropriate to be included in a real-time system to diagnose EDE.

Finally, the error/plot graphs for mC-ReliefF can be seen in Fig. 7 (top). In this case, and as can be confirmed by the statistical tests shown in Fig. 8, the cost significantly decreases as long as λ is greater than 5, whereas no significant differences were found in the case of the error.

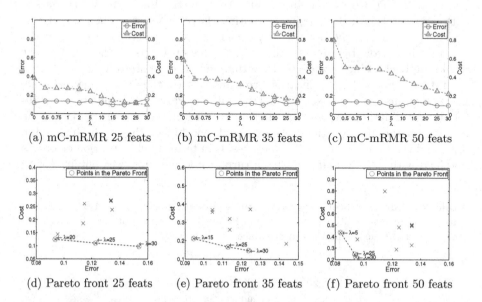

(a) mC-mRMR 25 feats   (b) mC-mRMR 35 feats   (c) mC-mRMR 50 feats

(d) Pareto front 25 feats   (e) Pareto front 35 feats   (f) Pareto front 50 feats

Fig. 5. Error/cost plots (top) and Pareto front (bottom) for mC-mRMR and different values of λ and selected features (25, 35 and 50) (Color figure online)

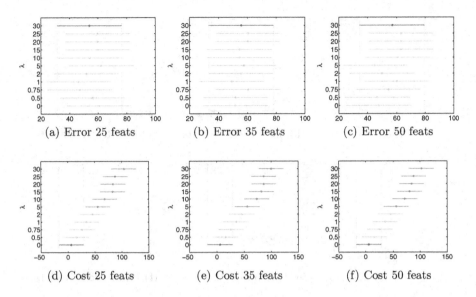

(a) Error 25 feats   (b) Error 35 feats   (c) Error 50 feats

(d) Cost 25 feats   (e) Cost 35 feats   (f) Cost 50 feats

Fig. 6. Statistical tests for mC-mRMR and different values of λ and selected features (25, 35 and 50)

Table 4. Mean classification error(%) and time (milliseconds) on average for the 10 folds for the Pareto front points obtained by mC-mRMR. Best error and time are marked in bold face.

Feats	λ	Error	Time
25	20	9.36	12.45
	25	12.27	9.51
	30	15.36	**8.42**
35	15	9.45	24.24
	25	11.27	17.66
	30	12.36	16.28
50	5	**8.36**	50.35
	25	9.36	25.24
	30	9.45	23.89

Figure 7 (bottom) and Table 5 present the results for those points in the Pareto front; i.e., minimizing either the cost or the error. The best result in terms of classification error (6.64) was obtained with $\lambda = 0$ when retaining 50 features per fold, improving previous results with the other feature selection methods. In turn, the lowest time was obtained when retaining 25 features per fold for whatever value of λ, but at the expense of increasing the error in at least 3.72 %. In this situation, and after consulting the clinical experts, the authors think that it is better to choose a trade-off between cost and error. The error obtained when $\lambda = 1$ and 35 features are retained is 7.55 %, which is slightly higher than

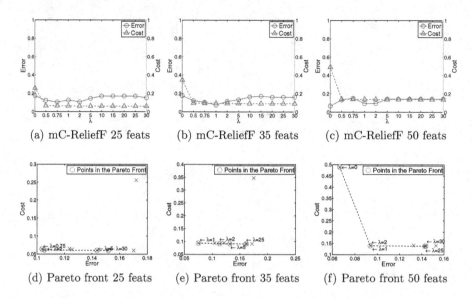

(a) mC-ReliefF 25 feats   (b) mC-ReliefF 35 feats   (c) mC-ReliefF 50 feats

(d) Pareto front 25 feats   (e) Pareto front 35 feats   (f) Pareto front 50 feats

Fig. 7. Error/cost plots (top) and Pareto front (bottom) for mC-ReliefF and different values of λ and selected features (25, 35 and 50)

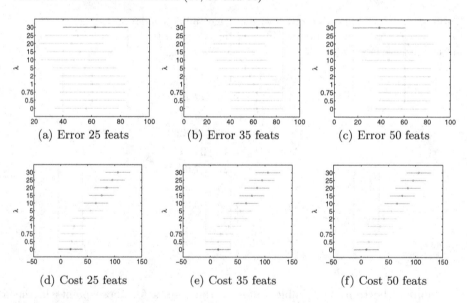

(a) Error 25 feats   (b) Error 35 feats   (c) Error 50 feats

(d) Cost 25 feats   (e) Cost 35 feats   (f) Cost 50 feats

Fig. 8. Statistical tests for mC-ReliefF and different values of λ and selected features (25, 35 and 50)

the best one but no statistical differences were found between them. With this combination the time required is 7.11 ms, which reduces the time significantly to a sixth fraction of the one obtained with the best error (47.98 ms).

Table 5. Mean classification error(%) and time (milliseconds) on average for the 10 folds for the Pareto front points obtained by mC-ReliefF. Best error and time are marked in bold face.

Feats	λ	Error	Time
25	0.75	10.36	**3.50**
	2	10.55	**3.50**
	5	14.36	**3.50**
	30	15.18	**3.50**
35	1	7.55	7.11
	2	11.36	7.04
	5	13.18	7.00
	25	16.09	7.00
50	0	**6.64**	47.98
	1	9.36	10.62
	2	9.36	10.62
	25	14.27	10.62
	30	14.36	10.62

The time required by previous approaches, which did not consider the use of feature selection, prevented their clinical use since they could not work in real time. Thus, the reduction in the computing time presented in this paper leads to a significant improvement, in such a way that the methodology for tear film classification could be used in the clinical practice to support dry eye diagnosis. Moreover, the creation of the tear film maps previously mentioned (see Sect. 1) could be manageable thanks to the optimization proposed in our research (only 7.11 ms to extract the features in each window), since the application of this method to thousands of local windows would not be a bottleneck anymore.

6   Conclusions

In this paper we have presented a real problem in which the cost associated with the features (in this case, in the form of extracting time) plays a crucial role. For the automatic classification of the tear film lipid layer, it is necessary to extract features from the eye images. However, the extraction step involves the computation of several statistics which is not an homogeneous operation, so the use of one or another feature makes a big difference in computational time. Using the whole set of extracted features is not feasible, since it prevents the use of the classification methodology in real time. Feature selection can be a solution to this problem but, unfortunately, the great majority of existing methods do not take the feature cost into account.

    Thus, in this paper we propose to apply a framework for feature selection that uses the cost of the features as part of the evaluation function to assess the goodness of a given feature or subsets of features, trying to achieve a trade-off between the relevance of a feature and its cost. Within this framework, moreover, it is possible to adjust the influence of the cost by means of a parameter named λ, allowing a fine tuning of the selection process balancing performance and cost according to the user's needs. Specifically, we have chosen three well-known feature selection methods: CFS, mRMR and ReliefF. The experimental results with a SVM classifier showed that it is possible to reduce considerably the cost without significantly increasing the classification error (compared with the original version of the feature selection methods, i.e., without taking the cost into account). The best result was obtained with the cost-based version of ReliefF, permitting to decrease the computational time to extract the features to only 7 ms, with a classification accuracy over 92 %. In clinical terms, the manual process done by experts can be automated with the benefits of being faster and unaffected by subjective factors. The clinical significance of the results should be highlighted, as the agreement between subjective observers is between 91 %–100 % according to [18].

As future work, we plan to introduce the cost function to more sophisticated feature selection methods, such as embedded or wrappers. It would be also interesting to test the proposed method on other real problems which also take into account the cost of the input features. Regarding tear film classification, and taking into account the real-time availability of the proposed method, our future research also includes to apply it at a local level in order to analyze heterogeneous tear film images.

Acknowledgments. This research has been partially funded by the Secretaría de Estado de Investigación of the Spanish Government and FEDER funds of the European Union through the research projects TIN2012-37954 and PI14/02161; and by the Consellería de Industria of the Xunta de Galicia through the research projects GPC2013/065 and GRC2014/035.

We would also like to thank the Optometry Service of the University of Santiago de Compostela (Spain) for providing us with the annotated dataset.

References

1. Pflugfelder, S., Tseng, S., Sanabria, O., Kell, H., Garcia, C., Felix, C., Feuer, W., Reis, B.: Evaluation of subjective assessments and objective diagnostic tests for diagnosing tear-film disorders known to cause ocular irritation. Cornea **17**(1), 38–56 (1998)
2. Wolff, E.: Anatomy of the Eye and Orbit, 4th edn. H.K. Lewis and Co., London (1954)
3. Rolando, M., Iester, M., Macrí, A., Calabria, G.: Low spatial-contrast sensitivity in dry eyes. Cornea **17**(4), 376–379 (1998)
4. Lemp, M.A., Baudouin, C., Baum, J., Dogru, M., Foulks, G.N., Kinoshita, S., Laibson, P., McCulley, J., Murube, J., Pfugfelder, S.C., Rolando, M., Toda, I.: The definition and classification of dry eye disease: report of the definition and classification subcommittee of the international dry eye workshop. Ocul. Surf. **5**(2), 75–92 (2007)

5. O'Brien, P.D., Collum, L.M.: Dry eye: diagnosis and current treatment strategies. Curr. Allergy Asthma Rep. **4**(4), 314–319 (2004)
6. Guillon, J.P., Guillon, M.: Tearscope Plus Clinical Handbook and Tearscope Plus Instructions. Keeler Ltd., Keeler Inc, Windsor, Broomall (1997)
7. Guillon, J.P.: Non-invasive tearscope plus routine for contact lens fitting. Contact Lens & Anterior Eye **21**(Suppl 1), 31–40 (1998)
8. Ramos, L., Penas, M., Remeseiro, B., Mosquera, A., Barreira, N., Yebra-Pimentel, E.: Texture and color analysis for the automatic classification of the eye lipid layer. In: Cabestany, J., Rojas, I., Joya, G. (eds.) IWANN 2011, Part II. LNCS, vol. 6692, pp. 66–73. Springer, Heidelberg (2011)
9. Remeseiro, B., Penas, M., Mosquera, A., Novo, J., Penedo, M.G., Yebra-Pimentel, E.: Statistical comparison of classifiers applied to the interferential tear film lipid layer automatic classification. Comput. Math. Methods Med. **2012**, 1–10 (2012)
10. Remeseiro, B., Penas, M., Barreira, N., Mosquera, A., Novo, J., García-Resúa, C.: Automatic classification of the interferential tear film lipid layer using colour texture analysis. Comput. Methods Programs Biomed. **111**, 93–103 (2013)
11. Remeseiro, B., Mosquera, A., Penedo, M.G., García-Resúa, C.: Tear film maps based on the lipid interference patterns. In: Proceedings of the 6th International Conference on Agents and Artificial Intelligence, pp. 732–739 (2014)
12. Remeseiro, B., Mosquera, A., Penedo, M.G.: CASDES: a computer-aided system to support dry eye diagnosis based on tear film maps. IEEE J. Biomed. Health Inform. (2015, in press)
13. Guyon, I.: Feature Extraction: Foundations and Applications, vol. 207. Springer, Heidelberg (2006)
14. Remeseiro, B., Bolon-Canedo, V., Peteiro-Barral, D., Alonso-Betanzos, A., Guijarro-Berdiñas, B., Mosquera, A., Penedo, M.G., Sánchez-Maroño, N.: A methodology for improving tear film lipid layer classification. IEEE J. Biomed. Health Inf. **18**(4), 1485–1493 (2014)
15. Bolón-Canedo, V., Porto-Díaz, I., Sánchez-Maroño, N., Alonso-Betanzos, A.: A framework for cost-based feature selection. Pattern Recogn. **47**(7), 2481–2489 (2014)
16. Bolón-Canedo, V., Remeseiro, B., Sánchez-Maroño, N., Alonso-Betanzos, A.: mC-ReliefF: an extension of relieff for cost-based feature selection. In: Proceedings of the 6th International Conference on Agents and Artificial Intelligence, pp. 42–51 (2014)
17. Rolando, M., Valente, C., Barabino, S.: New test to quantify lipid layer behavior in healthy subjects and patients with keratoconjunctivitis sicca. Cornea **27**(8), 866–870 (2008)
18. García-Resúa, C., Giráldez-Fernández, M.J., Penedo, M.G., Calvo, D., Penas, M., Yebra-Pimentel, E.: New software application for clarifying tear film lipid layer patterns. Cornea **32**(4), 536–546 (2013)
19. McLaren, K.: The development of the CIE 1976 (L*a*b) uniform colour-space and colour-difference formula. J. Soc. Dyers Colour. **92**(9), 338–341 (1976)
20. Haralick, R.M., Shanmugam, K., Dinstein, I.: Texture features for image classification. IEEE Trans. Syst. Man Cybern. **3**, 610–621 (1973)
21. Bradski, G.: The OpenCV Library. Dr. Dobb's J. Softw. Tools (2000)
22. Kotsiantis, S.B.: Supervised machine learning: a review of classification techniques. Informatica **31**, 249–268 (2007)
23. Burges, C.: A tutorial on support vector machines for pattern recognition. Data Min. Knowl. Disc. **2**(2), 1–47 (1998)

24. Yang, J., Honavar, V.: Feature subset selection using a genetic algorithm. IEEE Intell. Syst. Appl. **13**(2), 44–49 (1998)
25. Bahamonde, A., Bayón, G., Díez, J., Quevedo, J., Luaces, O., Del Coz, J., Alonso, J., Goyache, F.: Feature subset selection for learning preferences: a case study. In: Proceedings of the Twenty-first International Conference on Machine Learning, pp. 49–56. ACM (2004)
26. Haralick, R., Shanmugam, K., Dinstein, I.: Textural features for image classification. IEEE Trans. Syst. Man Cybern. **SMC–3**(6), 610–621 (1973)
27. Feddema, J., Lee, C., Mitchell, O.: Weighted selection of image features for resolved rate visual feedback control. IEEE Trans. Robot. Autom. **7**(1), 31–47 (1991)
28. Friedman, J.: Regularized discriminant analysis. J. Am. Stat. Assoc. **84**(405), 165–175 (1989)
29. Wright, J., Yang, A., Ganesh, A., Sastry, S., Ma, Y.: Robust face recognition via sparse representation. IEEE Trans. Pattern Anal. Mach. Intell. **31**(2), 210–227 (2009)
30. You, D., Hamsici, O., Martinez, A.: Kernel optimization in discriminant analysis. IEEE Trans. Pattern Anal. Mach. Intell. **33**(3), 631–638 (2011)
31. Xu, Z., Kusner, M., Weinberger, K., Chen, M., Chapelle, O.: Classifier cascades and trees for minimizing feature evaluation cost. J. Mach. Learn. Res. **15**(1), 2113–2144 (2014)
32. Huang, C., Wang, C.: A ga-based feature selection and parameters optimization for support vector machines. Expert Syst. Appl. **31**(2), 231–240 (2006)
33. Sivagaminathan, R., Ramakrishnan, S.: A hybrid approach for feature subset selection using neural networks and ant colony optimization. Expert Syst. Appl. **33**(1), 49–60 (2007)
34. Jiawei, H., Kamber, M.: Data Mining: Concepts and Techniques. Morgan Kaufmann, San Francisco (2001)
35. Hall, M.A.: Correlation-based feature selection for machine learning. PhD thesis, The University of Waikato (1999)
36. Rich, E., Knight, K.: Artificial Intelligence. McGraw-Hill, New York (1991)
37. Peng, H., Long, F., Ding, C.: Feature selection based on mutual information criteria of max-dependency, max-relevance, and min-redundancy. IEEE Trans. Pattern Anal. Mach. Intell. **27**(8), 1226–1238 (2005)
38. Kononenko, I.: Estimating attributes: analysis and extensions of RELIEF. In: Bergadano, F., De Raedt, L. (eds.) ECML 1994. LNCS, vol. 784, pp. 171–182. Springer, Heidelberg (1994)
39. Kira, K., Rendell, L.A.: A practical approach to feature selection. In: Proceedings of the Ninth International Workshop on Machine Learning, pp. 249–256. Morgan Kaufmann Publishers Inc. (1992)
40. VARPA: VOPTICAL_I1, VARPA optical dataset annotated by optometrists from the Optometry Service, University of Santiago de Compostela (Spain) (2012). http://www.varpa.es/voptical_I1.html. Accessed July 2015
41. Hochberg, Y., Tamhane, A.C.: Multiple Comparison Procedures. Wiley, Hoboken (1987)
42. Teich, J.: Pareto-front exploration with uncertain objectives. In: Zitzler, E., Thiele, L., Deb, K., Coello, C.A.C., Corne, D. (eds.) Evolutionary Multi-criterion Optimization. LNCS, vol. 1993, pp. 314–328. Springer, Heidelberg (2001)

Scalarized and Pareto Knowledge Gradient for Multi-objective Multi-armed Bandits

Saba Yahyaa$^{(\boxtimes)}$, Madalina M. Drugan, and Bernard Manderick

Computer Science Department, Vrije Universiteit Brussel,
Pleinlaan 2, 1050 Brussels, Belgium
{syahyaa,mdrugan,bmanderi}@vub.ac.be
https://ai.vub.ac.be

Abstract. A multi-objective multi-armed bandit (MOMAB) problem is a sequential decision process with stochastic reward vectors. We extend knowledge gradient (KG) policy to the MOMAB problem, and we propose Pareto-KG and scalarized-KG algorithms. The Pareto-KG trades off between exploration and exploitation by combining KG policy with Pareto dominance relations. The scalarized-KG makes use of a linear or non-linear scalarization function to convert the MOMAB problem into a single-objective multi-armed bandit problem and uses KG policy to trade off between exploration and exploitation. To measure the performance of the proposed algorithms, we introduce three regret measures. We compare empirically the performance of the KG policy with UCB1 policy on a test suite of MOMAB problems with normal distributions. The Pareto-KG and scalarized-KG are the algorithms with the best empirical performance.

Keywords: Multi-armed bandit problems · Multi-objective optimization · Exploration/exploitation · Knowledge gradient policy

1 Introduction

The standard multi-armed bandit (MAB) problem is a sequential stochastic problem of an agent that optimizes its decisions while improving its knowledge on arms. At each time step t, the agent pulls one arm a and receives reward signal. The reward is drawn from a stationary distribution, e.g. a normal distribution $N(\mu_a, \sigma_a^2)$, where μ_a is the mean and σ_a^2 is the variance of the reward distribution of the arm a. We assume that the mean and variance parameters are unknown to the agent. By drawing each arm a, the agent maintains estimations of the true mean and variance which are known as $\hat{\mu}_a$ and $\hat{\sigma}_a^2$, respectively. The agent's goal is to minimize the *loss* of not pulling the optimal arm a^* that has the maximum mean all the time. For number of time steps L, the loss (the *total expected regret*) R_L is $R_L = L\mu^* - \sum_{t=1}^{L} \mu_t$, where $\mu^* = \max_{a=1,\cdots,|A|} \mu_a$ is the true mean of the optimal arm a^* and μ_t is the true mean of the selected arm a at time step t.

At each time step t, the agent either selects the arm that has the maximum estimated mean $\hat{\mu}^*$ (exploitation of the greedy arm), or selects one of the other

© Springer-Verlag Berlin Heidelberg 2015
N.T. Nguyen et al. (Eds.): TCCI XX, LNCS 9420, pp. 99–116, 2015.
DOI: 10.1007/978-3-319-27543-7_5

arms in order to improve the corresponding estimate parameters of the reward distribution (exploration of the other arms). The agent has to find a proper *trade-off between exploitation and exploration* (or trade-off for short) [1] to minimize the total expected regret R_L of not pulling the optimal arm. To find a good trade-off, [2] compared several action selection policies on the MAB problems and showed that Knowledge Gradient (KG) policy [3] outperforms other MAB policies including the Upper Confidence Bound (UCB1) policy [4], although, it does not have any parameters to be tuned. UCB1 and KG have are similar trade off between exploitation and exploration. Both policies add an exploration bound $ExpB_a$ to the estimated mean of each arm a and select the arm that has the highest combined value of estimated mean and exploration bound. But in case of UCB1, the exploration bound of an arm a requires only knowledge about that arm itself, while in case of KG it also requires knowledge about the other arms.

The multi-objective MAB (MOMAB) problem is a MAB problem [5], where the agent selects an arm a, the agent receives a reward vector r_a instead of a single scalar value. The reward vector of the arm a is independent from all other arms and drawn from a corresponding stationary probability distribution vector with unknown parameters, e.g. from a multivariate normal distribution $N(\mu_a, \sigma_a^2)$, where μ_a is the mean vector and σ_a^2 is the covariance matrix of the arm a. The mean vector μ_a of arm a has independent D dimensions (objectives). Moreover, the D objectives are conflicting with each others. The agent has to trade off the conflicting objectives of the mean vectors, since there is not only one optimal arm instead there is a set of optimal arms (Pareto front A^*). The goal of the agent is to find out the Pareto front (exploring the optimal arms) to minimize the total Pareto loss and play fairly the optimal arms in the Pareto front (exploiting the optimal arms) to minimize the total unfairness loss [6, 7]. At each time step t, the Pareto loss (Pareto regret) is the distance between the mean vector set of the Pareto front and the mean vector of the selected arm. The unfairness loss (unfairness regret) is the Shannon entropy which is the measure of disarray on the frequency of selecting the optimal arms in the Pareto front A^*. The higher entropy is the higher disorder. The total Pareto and unfairness regrets are the cumulative summation of the Pareto and unfairness regrets over t time steps, respectively [8].

The Pareto front A^* is found either by using: (1) Pareto partial order relation (or Pareto dominance relation) [9], or (2) scalarized functions [10]. The Pareto dominance relation finds the Pareto front A^* by optimizing directly the Multi-Objective (MO) space. The scalarized functions convert the MO space to a single-objective space, i.e. the mean vectors are transformed into scalar values. There are two types of scalarization functions, linear and non-linear (e.g., Chebyshev) functions. Linear scalarized function is simple but can not find all the optimal arms in a non-convex mean vectors set. In opposition, Chebyshev scalarized function has an extra parameter to be tuned, however, it can find all the optimal arms in a non-convex mean vectors set. Recently, [5] introduced a MO version of the Upper Confidence Bound(UCB1) policy to improve the performance of the Pareto dominance relation and scalarized functions.

In this paper, we extend KG policy [3] to the MOMAB problem. Pareto KG combines KG policy with the Pareto dominance relation and scalarized functions KG combine KG policy with either the linear or Chebychev scalarized function. Note that, the authors in [4,11] introduced other versions of UCB1 policy for normal reward distribution. In this paper, we use the same version of Bernoulli reward distributions as [12].

The rest of the paper is organized as follows. Section 2 presents background information on the algorithms and the used notation. Section 3 introduces UCB1 in MO normal distributions bandits. Section 4 discusses KG policy and proposes Pareto-KG algorithm, two variants of the linear scalarized KG functions: linear scalarized-KG across arms and objectives algorithms, and Chebyshev scalarized KG algorithm. Section 5 describes the experiments set up followed by experimental results where we compare UCB1 and KG on a test suite of MOMAB problems. Section 6 concludes and discusses future work.

2 Background

In this section, we introduce MOMAB framework, the Pareto dominance relation, scalarization functions, and regret measures of the MOMAB problems.

2.1 MOMAB Framework

Let us consider the MOMAB problems with $|A| \geq 2$ *independent* arms and *independent* D objectives per arm. At each time step t, an agent pulls one arm $a \in A$ and receives a reward vector $\boldsymbol{r}_a = [r_a^1, \cdots, r_a^d, \cdots, r_a^D]^T$, where T is the transpose. The reward vector $\boldsymbol{r}_a$ of an arm a is drawn from a multivariate normal distribution $N(\boldsymbol{\mu}_a, \boldsymbol{\sigma}_a^2)$, where $\boldsymbol{\mu}_a$ is the mean vector and $\boldsymbol{\sigma}_a^2$ is the diagonal covariance matrix or variance vector. We assume that the mean and the variance vector of each arm are *unknown* parameters to the agent. By drawing each arm a, the agent estimates the mean $\hat{\mu}_a^d$ and the variance $\hat{\sigma}_a^{2,\,d}$ parameters in each objective $d \in D$. The agent updates the estimated parameters by using either frequentist or Bayesian update rule [13]. Since the frequentist update rule is easier when both the mean and the variance vector of the reward distributions are unknown to the agent, we will use it. If at time step t, arm a is selected and reward vector $\boldsymbol{r}_{t+1}$ is obtained, the estimated mean $\hat{\mu}_a^d$ and the estimated variance $\hat{\sigma}_a^{2,\,d}$ of the arm a in each objective d are updated recursively as [13]:

$$N_a \leftarrow N_a + 1 \tag{1}$$

$$\hat{\sigma}_a^{2,\,d} \leftarrow \frac{N_a - 2}{N_a - 1} \hat{\sigma}_a^{2,\,d} + \frac{1}{N_a}(r_{t+1}^d - \hat{\mu}_a^d)^2, \quad \hat{\mu}_a^d \leftarrow (1 - \frac{1}{N_a})\,\hat{\mu}_a^d + \frac{1}{N_a}r_{t+1}^d$$

where N_a is the number of times arm a has been selected and r_{t+1}^d is the collected reward from the arm a in the objective d. Note that the variance has to be updated before the mean since it uses the old mean.

The mean $\boldsymbol{\mu}_a$ and standard deviation $\boldsymbol{\sigma}_a$ vector of each arm a are represented as $\boldsymbol{\mu}_a = [\mu_a^1, \cdots, \mu_a^D]^T$, and $\boldsymbol{\sigma}_a = [\sigma_a^1, \cdots, \sigma_a^D]^T$, respectively. When the objectives are conflicting with one another then the mean μ_a^d of arm a corresponding with objective d, can be better than the component $\mu_{a'}^d$ of another arm a' but worse if we compare the components for another objective $d' : \mu_a^d \succ \mu_{a'}^d$ but $\mu_a^{d'} \prec \mu_{a'}^{d'}$ for objectives d and d', respectively. The agent has a set of optimal arms (Pareto front A^*) which can be found either by Pareto dominance relation, or scalarization functions. The goal of the agent is to trade off between exploration (i.e., finding the Pareto front A^*) and exploitation (i.e., playing fairly the optimal arms in the set A^*) [6].

2.2   The Pareto Dominance Relation

Pareto dominance relation finds the Pareto front A^* directly in the MO space [9]. It uses the following relations between the mean vectors $\boldsymbol{\mu}_a$, and $\boldsymbol{\mu}_{a'}$ of arms a, and a', respectively: (1) Arm a dominates or is better than a', $a \succ a'$, if there exists at least one objective d for which $\mu_a^d \succ \mu_{a'}^d$ and for all other objectives d' we have $\mu_a^{d'} \succeq \mu_{a'}^{d'}$. (2) Arm a is incomparable with a', $a \parallel a'$, if and only if there exists at least one objective d for which $\mu_a^d \succ \mu_{a'}^d$ and there exists another objective d' for which $\mu_a^{d'} \prec \mu_{a'}^{d'}$. (3) Arm a is not dominated by a', $a' \nsucc a$, if and only if there exists at least one objective d for which $\mu_{a'}^d \prec \mu_a^d$. This means that either $a \succ a'$, or $a \parallel a'$. Using the above relations, the Pareto front A^* contains the optimal arms that are not dominated by all other arms. Moreover, the optimal arms in the A^* are incomparable with each other.

2.3   The Scalarization Functions

In general, scalarization functions convert the MO optimization problem into a single-objective one [10]. However, solving a MO optimization problem means finding the Pareto front A^*. We need a set of scalarized functions $\boldsymbol{F} = \{f^1, \cdots, f^s, \cdots, f^S\}$ to generate a variety of elements belonging to the Pareto front A^*. Each scalarized function $f^s \in \boldsymbol{F}$ has a corresponding predefined weight set $\boldsymbol{w}^s \in \boldsymbol{W}$, where $\boldsymbol{W} = (\boldsymbol{w}^1, \cdots, \boldsymbol{w}^S)$ is the total weight set. The total weight set $\boldsymbol{W}$ is uniformly random spread sampling in the weighted space [14]. There are two types of scalarization functions that weigh the mean vector, linear and non-linear (e.g., Chebyshev) scalarization functions.

The linear scalarized function assigns to each value of the mean vector of an arm a a weight w^d and the result is the sum of these weighted mean values. Given a predefined set of weights $\boldsymbol{w}^s = (w^1, \cdots, w^D)$ such that $\sum_{d=1}^{D} w^d = 1$, the linear scalarized function over the mean vector $\boldsymbol{\mu}_a$ of the arm a is:

$$f^s(\boldsymbol{\mu}_a) = w^1 \mu_a^1 + \cdots + w^D \mu_a^D \tag{2}$$

where $f^s(\boldsymbol{\mu}_a)$ is a linear scalarized function $s \in S$. The linear scalarization is intuitive and very popular because of its simplicity. However, it can not find all the optimal arms in the Pareto front A^* when the Pareto front is a *non-convex* mean vector set [14].

The Chebyshev scalarized function in addition to weights, Chebyshev scalarized function has a D-dimensional reference point, i.e. $z = [z^1, \cdots, z^D]^T$. It can find all the arms in a non-convex Pareto mean front set by moving the reference point [15]. For maximization MOMAB problems, the Chebyshev scalarized is [5]:

$$f^s(\boldsymbol{\mu}_a) = \min_{1 \leq d \leq D} w^d(\mu_a^d - z^d), \ \forall_a, \quad \text{where } z^d = \min_{1 \leq a \leq A} \mu_a^d - \epsilon^d, \ \forall_d, \quad (3)$$

and $\epsilon > 0$ is a small value. The reference point z is dominated by all the optimal mean vectors. It is the minimum of the current mean vector minus ϵ value.

After transforming the MO problem to a single-objective one, the scalarized functions f^s select the arm $a_{f^s}^*$ that has the maximum scalarized value:

$$a_{f^s}^* = \underset{1 \leq a \leq A}{\operatorname{argmax}} f^s(\boldsymbol{\mu}_a)$$

2.4   The Regret Metrics

To measure the performance of the Pareto dominance relation and the scalarization functions, the authors in [5,8] proposed three regret metric criteria.

Pareto regret metric (R_P) measures the distance between the mean vector of an arm a that is pulled at time step t and the Pareto front A^* mean vectors [5]. The R_P is calculated by finding firstly the virtual distance dis^*. At time step t, the virtual distance dis^* is defined as the minimum distance that is added to the mean vector $\boldsymbol{\mu}_t$ of the pulled arm a_t in each objective to create a virtual mean vector $\boldsymbol{\mu}_t^*$, $\boldsymbol{\mu}_t^* = \boldsymbol{\mu}_t + \boldsymbol{\varepsilon}^*$ that is incomparable $(\boldsymbol{\mu}_t^* || \boldsymbol{\mu}_{a^*} \ \forall_{a^* \in A^*})$ with all the mean vectors of the optimal arms a^* in the Pareto front A^*, where $\boldsymbol{\varepsilon}^* = [dis^{*,1}, \cdots, dis^{*,D}]^T$ is a vector. The Pareto regret R_P is $R_P = dis(\boldsymbol{\mu}_t, \boldsymbol{\mu}_t^*) = dis(\boldsymbol{\varepsilon}^*, \mathbf{0})$, where $dis(\boldsymbol{\mu}_t, \boldsymbol{\mu}_t^*) = (\sum_{d=1}^{D}(\mu_t^{*,d} - \mu_t^d)^2)^{0.5}$ is the Euclidean distance between the mean vector of the virtual arm $\boldsymbol{\mu}_t^*$ and the mean vector of the pulled arm $\boldsymbol{\mu}_t$ at time step t, and $\mathbf{0}$ is a zero vector with D objectives. The regret of the Pareto front is 0 for optimal arms, the mean of the optimal arm coincides itself.

The scalarized regret metric (R_S) measures the distance between the maximum value of a scalarized function and the scalarized value of an arm that is pulled at time step t [5]. The $R_S(t) = \max_{1 \leq a \leq A} f^s(\boldsymbol{\mu}_a) - f^s(\boldsymbol{\mu}_{a'})(t)$ is the difference between the maximum value for a scalarized function f^s which is either Chebyshev or linear on the set of arms A and the scalarized value for an arm a' that is pulled by the scalarized f^s at time step t.

The unfairness regret metric (R_F) is Shannon's entropy [8]. It is a measure of disorder on the frequency of selecting the optimal arms in the Pareto front A^*. The higher the entropy, the higher the disorder. At time step t, the $R_F(t) = (1/N_{|A^*|}) \sum_{a^* \in A^*} p_{a^*}(t) \ln(p_{a^*}(t))$, where $p_{a^*}(t) = N_{a^*}(t)/N(t)$ is the probability of selecting an optimal arm a^* at time step t, where $N_{a^*}(t)$

is the number of times the optimal arm a^* has been selected, and $N(t)$ is the number of times all arms $a = 1, \cdots , |A|$ have been selected at time step t, and $N_{|A^*|}(t)$ is the number of times the optimal arms, $a^* = 1, \cdots , |A^*|$ have been selected at time step t.

Note that, the authors in [5,16] used another version of the unfairness regret metric which is related to the variance in drawing all the optimal arms in the Pareto front A^*. However, the unfairness regret grows exponentially on the number of time steps t and does not take into its account the total number of selecting the optimal arms in A^*.

3   UCB1 for MOMAB Problems

In MOMAB problems, UCB1 policy [5] plays firstly each arm a once, and adds to the estimated mean $\hat{\mu}_a$ of each arm a a corresponding exploration bound ExpB_a. The exploration bound is an upper confidence bound which depends on the number of times arm a has been selected. In this paper, we use UCB1 in the MOMAB problems with normal distributions.

3.1   Pareto-UCB1 for Normal MOMAB Problem

Pareto-UCB1 plays initially each arm a once. At each time step t, it estimates the mean vector $\hat{\boldsymbol{\mu}}_a$ of each of the MO arms a, i.e. $\hat{\boldsymbol{\mu}}_a = [\hat{\mu}_a^1, \cdots , \hat{\mu}_a^D]^T$ and adds to each objective $d \in D$ an exploration bound ExpB_a^d to trade off between the exploration and exploitation. The exploration bound $\mathrm{ExpB}_a^d = \sqrt{(2 \ln(t \sqrt[4]{D|A^*|}))/N_a}$ is the upper confidence bound of the estimated mean of the arm a in the objective d, where N_a is the number of times arm a has been selected, $|A^*|$ is the number of optimal arms and D is the number of objectives. Pareto-UCB1 uses Pareto dominance relation, c.f. Sect. 2.2 to find its Pareto front A^*_{UCB1}. For all the non-optimal arms $k \notin A^*_{UCB1}$ there exists a Pareto optimal arm $j \in A^*_{UCB1}$ that is not dominated by the arms k such that: $\hat{\boldsymbol{\mu}}_k + \mathbf{ExpB}_k \not\succ \hat{\boldsymbol{\mu}}_j + \mathbf{ExpB}_j$, where $\mathbf{ExpB}_a = [\mathrm{ExpB}_a^1, \cdots , \mathrm{ExpB}_a^D]^T$ is the exploration bound vector of the arm a with $\mathrm{ExpB}_a^1 = \cdots = \mathrm{ExpB}_a^D$. Pareto-UCB1 selects uniformly at random one of the arms in the set A^*_{UCB1}. The idea is to select most of the times one of the optimal arm in the Pareto front. An arm $a' \notin A^*$ that is closer to the Pareto front according to metric measure is more selected than the arm $k \notin A^*$ that is far from A^*.

3.2   Scalarized-UCB1 for Normal MOMAB Problem

Scalarized-UCB1 adds an upper confidence bound to the pulled arm under the scalarized function f^s. Each scalarized function f^s has an associated predefined set of weights, $(w^1, \cdots , w^D)^s$, $\sum_{d=1}^D w^d = 1$. The upper bound depends on the number of times N^s the scalarized function f^s has been selected, and on the number of times N_a^s the arm a has been pulled under the scalarized function f^s.

We use s to refer the scalarized function f^s. Firstly, the scalarized UCB1 plays each arm once and estimates the mean vector $\hat{\boldsymbol{\mu}}_a$ of each arm $a \in A$. At each time step t, it selects one scalarized function f^s uniformly at random and pulls the optimal arm a^*, the arm that has the maximum scalarized value plus the exploration bound as:

$$a^* = \max_{1 \leq a \leq A} (f^s(\hat{\boldsymbol{\mu}}_a) + \sqrt{\frac{2\ln(N^s)}{N_a^s}})$$

where f^s is either linear scalarized function as can be seen in Eq. (2), or Chebyshev scalarized function, see Eq. (3) with a predefined weight set.

4   Multi Objective Knowledge Gradient

Knowledge gradient (KG) policy [3] is an index policy that determines for arm a the index V_a^{KG} as:

$$V_a^{KG} = \hat{\bar{\sigma}}_a * x \left(-\left| \frac{\hat{\mu}_a - \max_{a' \neq a,\, a' \in |A|} \hat{\mu}_{a'}}{\hat{\bar{\sigma}}_a} \right| \right)$$

where $\hat{\mu}_a$ is the estimated mean of the reward distribution an arm a, and $\hat{\bar{\sigma}}_a = \hat{\sigma}_a/N_a$ is the Root Mean Square Error (RMSE) of the estimated mean of an arm a, where $\hat{\sigma}_a^2$ is the estimated variance of the reward distribution an arm a, and N_a is the number of times arm a has been pulled. The function $x(\zeta) = \zeta\Phi(\zeta) + \phi(\zeta)$ where $\phi(\zeta) = 1/\sqrt{2\pi}\exp(-\zeta^2/2)$ is the standard normal density $N(0,1)$ and its cumulative distribution is $\Phi(\zeta) = \int_{-\infty}^{\zeta} \phi(\zeta')d\zeta'$.

The KG policy chooses the arm a with the largest V_a^{KG} and it prefers those arms about which comparatively little is known. These arms are the ones whose distributions around the estimate mean, $\hat{\mu}_a$ have larger estimated standard deviations, $\hat{\sigma}_a$. The KG prefers an arm a over its alternatives if its confidence in the estimate mean $\hat{\mu}_a$ is low. This policy trades off between exploration and exploitation by selecting its arm a_{KG}^* as:

$$a_{KG}^* = \underset{a \in A}{\mathrm{argmax}} \left(\hat{\mu}_a + (L - t) V_a^{KG} \right) \tag{4}$$

where t is a time step and L is the horizon of an experiment which is the total number of plays that the agent has. In [2], KG policy is the competitive policy for the standard multi-armed bandit problems according to the collected cumulated average reward and average frequency of optimal selection performances. Moreover, KG policy does not have any parameter to be tuned. Therefore, we used KG policy in the MOMAB problems.

Note that, [3] used the change in the standard deviation $\Delta\sigma = (\frac{\hat{\bar{\sigma}}^2}{1+\sigma^2/\hat{\bar{\sigma}}^2})^{0.5}$ instead of the RMSE. Since, the variance of the reward distribution σ^2 is unknown parameter to the agent and we use the frequentist view to update the estimated parameters (i.e., the estimated variance $\hat{\sigma}^2$ and the estimated mean $\hat{\mu}$ of the normal distributed reward $N(\mu, \sigma^2)$) we use the RMSE as [2]. The authors in [2] used a version of KG which combines feature from the KG policy and the expected improvement policy [11]. The version of KG requires simpler computation than KG, although it performs as well as KG policy.

4.1 Pareto Knowledge Gradient Algorithm

Pareto KG algorithm uses Pareto dominance relation [9] (Sect. 2.2) to order arms and KG policy to trade off between finding and playing evenly the optimal arms. The pseudocode of the Pareto-KG is given in Fig. 1. At each time step t, Pareto-KG computes for each arm $a \in A$, the corresponding exploration bound vector $\mathbf{ExpB}_a = [\text{ExpB}_a^1, \cdots, \text{ExpB}_a^D]^T$. The exploration bound ExpB_a^d of an arm a in the objective d depends on the estimated mean $\hat{\mu}_a^d$ of all arms $a = 1, \cdots A$ in that objective d and on the estimated standard deviation $\hat{\sigma}_a^d$ of the arm a in the objective d. It is calculated as:

$$\text{ExpB}_a^d = (L - t) * |A|D * v_a^d, \quad \text{where} \quad v_a^d = \hat{\bar{\sigma}}_a^d \; x(-|\frac{\hat{\mu}_a^d - \max\limits_{a' \neq a, \, a' \in A} \hat{\mu}_{a'}^d}{\hat{\bar{\sigma}}_a^d}|) \; \forall_d \in D$$

where v_a^d is the index of an arm a in the objective d, L is the horizon of an experiment which is the total number of time steps, $|A|$ is the total number of arms, D is the number of objectives and $\hat{\bar{\sigma}}_a^d$ is the RMSE of the arm in the objective d which equals $\hat{\sigma}_a^d / \sqrt{N_a}$. The N_a is the number of times the arm a has been pulled. After computing the exploration bound vector for each arm, Pareto-KG adds the exploration bound vector $\mathbf{ExpB}_a$ of arm a to the corresponding estimated mean vector $\hat{\boldsymbol{\mu}}_a$. Pareto-KG selects its optimal arms a^* that are not dominated by all other arms a, $a \in A$ (Step 4). Pareto-KG chooses uniformly at random one of the optimal arms in A_{KG}^* (Step 5). A_{KG}^* is the Pareto front KG, the set that contains Pareto optimal arms using KG policy. After pulling the chosen arm a^*, Pareto-KG algorithm observes the reward vector $\boldsymbol{r}_{a^*}$ (Step 6) updates the corresponding estimated mean $\hat{\boldsymbol{\mu}}_{a^*}$ vector, the estimated variance $\hat{\boldsymbol{\sigma}}_{a^*}^2$ vector, the number of times arm a^* is chosen N_{a^*} (Step 7), c.f. Eq. (1) and computes the Pareto R_P and the unfairness R_F regrets (Step 8).

4.2 Scalarized Knowledge Gradient Algorithm

Scalarized KG functions convert the MOMAB problem into a signal-objective MAB problem and make use of the estimated mean and variance. Scalarized KG functions are three variants, linear scalarized KG across arms and objectives and Chebyshev scalarized KG.

Linear scalarized-KG across arms (LS_1-KG) converts immediately the MOMAB problem into a single-objective MAB problem. It converts the estimated mean vector $\hat{\boldsymbol{\mu}}_a = [\hat{\mu}_a^1, \cdots, \hat{\mu}_a^D]^T$ and the estimated variance $\hat{\boldsymbol{\sigma}}_a^2 = [\hat{\sigma}_a^{2, \, 1}, \cdots, \hat{\sigma}_a^{2, \, D}]^T$ of each arm a into one-objective values and computes for each arm

1. **Input:** Horizon of an experiment L;time step t;number of arms $|A|$;number of objectives D;reward distribution $r_a \sim N(\mu_a, \sigma_a^2) \; \forall_a \in A$.
2. **Initialize:** Plays each arm a, *Initial* steps to estimate the mean vectors $\hat{\mu}_a = [\hat{\mu}_a^1, \cdots, \hat{\mu}_a^D]^T$ and the standard deviation vectors $\hat{\sigma}_a = [\hat{\sigma}_a^1, \cdots, \hat{\sigma}_a^D]^T$.
3. **For time step** $t = 1$ to L
4.   Find A_{KG}^* such that $\forall_{a^*} \in A_{KG}^*$ and $\forall_a \notin A_{KG}^*$
$$\hat{\mu}_a + \mathbf{ExpB}_a \; \not\succ \; \hat{\mu}_{a^*} + \mathbf{ExpB}_{a^*}$$
5.   Select a^* uniformly at random from A_{KG}^*
6.   Observe: reward vector r_{a^*}, $r_{a^*} = [r_{a^*}^1, \cdots, r_{a^*}^D]^T$
7.   Update: the mean vector $\hat{\mu}_{a^*}$;the variance vector $\hat{\sigma}_{a^*}^2$;$N_{a^*} \leftarrow N_{a^*} + 1$
8.   Compute: the unfairness regret;Pareto regret
9. **End for**
10. **Output:** Unfairness regret; Pareto regret.

Fig. 1. Algorithm: (Pareto-KG).

a the corresponding exploration bound ExpB_a to trade-off between the exploration and exploitation. At each time step t, LS_1-KG weighs both the estimated mean vector $\hat{\mu}_a$, and estimated variance vector $\hat{\sigma}_a^2$ of each arm a, converts the MO vectors to scalar values by summing the weighted elements of each vector. Thus, we have one-objective MAB problem. The KG calculates for each arm a, an exploration bound ExpB_a which depends on all other arms and selects the arm that has the maximum estimated mean plus exploration bound. The LS_1-KG is as:

$$\widetilde{\mu_a} = f^s(\hat{\mu}_a) = w^1 \hat{\mu}_a^1 + \cdots + w^D \hat{\mu}_a^D \qquad \forall_a \in A \tag{5}$$

$$\widetilde{\sigma}_a^2 = f^s(\hat{\sigma}_a^2) = w^1 \hat{\sigma}_a^{2,1} + \cdots + w^D \hat{\sigma}_a^{2,D} \qquad \forall_a \in A \quad \text{where} \tag{6}$$

$$\widetilde{\widetilde{\sigma}}_a^2 = \frac{\widetilde{\sigma}_a^2}{N_a} \; \forall_a \in A, \quad v_a = \widetilde{\widetilde{\sigma}}_a \; x(-|\frac{\widetilde{\mu_a} - \max\limits_{a' \neq a,\, a' \in A} \widetilde{\mu_{a'}}}{\widetilde{\widetilde{\sigma}}_a}|) \quad \forall_a \in A, \tag{7}$$

f^s is a linear scalarized function that has a corresponding predefined set of weight $(w^1, \cdots, w^D)$, $\widetilde{\mu_a}$, $\widetilde{\sigma}_a^2$ are the modified estimated mean and variance of an arm a, respectively which are scalar values and $\widetilde{\widetilde{\sigma}}_a^2$ is the modified RMSE of the arm a which is a scalar value. The v_a is the KG index of the arm a. The function $x(\zeta) = \zeta \Phi(\zeta) + \phi(\zeta)$ where Φ and ϕ are the cumulative distribution and the density of the standard normal distribution $N(0, 1)$, respectively. At each time step t, LS_1-KG selects the optimal arm a^* according to:

$$a_{LS_1KG}^* = \underset{a=1,\cdots,|A|}{\mathrm{argmax}} \; (\widetilde{\mu_a} + \mathrm{ExpB}) = \underset{a}{\mathrm{argmax}} \; \underset{a=1,\cdots,|A|}{} (\widetilde{\mu_a} + (L - t) * |A|D * v_a) \tag{8}$$

where ExpB_a is the exploration bound of arm a, D is the number of objective, $|A|$ is the number of arms, and L is the horizon of an experiment.

Linear scalarized-KG across objectives (LS_2-KG) computes the exploration bound $\mathbf{ExpB}_a = [\mathrm{ExpB}_a^1, \cdots, \mathrm{ExpB}_a^D]$ of each arm a, adds it to the corresponding estimated mean vector $\hat{\mu}_a$, and converts the MO problem to a single-objective one. At each time step t, LS_2-KG computes the exploration bounds for all objectives of each arm, sums the estimated mean in each objective with its corresponding exploration bound, weighs each objective and converts the MO into a single-objective value by summing the weighted objective. LS_2-KG is as:

$$f^s(\hat{\mu}_a) = w^1(\hat{\mu}_a^1 + \underset{a}{\text{ExpB}}) + \cdots + w^D(\hat{\mu}_a^D + \underset{a}{\text{ExpB}}) \ \forall_a \in A, \quad \text{where} \quad (9)$$

$$\underset{a}{\text{ExpB}} = (L - t) \times |A|D \times v_a^d, \quad v_a^d = \hat{\hat{\sigma}}_a^d \ x(-|\frac{\hat{\mu}_a^d - \underset{a' \neq a, \, a' \in A}{\max} \hat{\mu}_{a'}^d}{\hat{\hat{\sigma}}_a^d}|) \quad \forall_d \in D,$$

and v_a^d is the index, $\hat{\mu}_a^d$ is the estimated mean, $\hat{\hat{\sigma}}_a^d$ is the RMSE, and $ExpB_a^d$ is the exploration bound of the arm a in the objective d. The LS_2-KG selects the optimal arm a^* that has maximum $f^s(\hat{\mu}_a)$ as:

$$a^*_{LS_2KG} = \underset{a=1,\cdots,|A|}{\text{argmax}} \ f^s(\hat{\mu}_a) \quad (10)$$

Chebyshev Scalarized-KG (*Cheb*-KG) computes the exploration bound $ExpB_a^d$ of each arm a in each objective d and converts the MO problem into a single-objective one. The *Cheb*-KG is as:

$$f^s(\hat{\boldsymbol{\mu}}_a) = \underset{1 \leq d \leq D}{\min} w^d(\hat{\mu}_a^d + \underset{a}{\text{ExpB}} - z^d) \quad \forall_a \in A \quad (11)$$

where f^s is a Chebyshev scalarized function that has a predefined set of weights $(w^1, \cdots, w^D)$. The exploration bound $ExpB_a^d$ is calculated as:

$$\underset{a}{\text{ExpB}} = (L - t) * |A|D * v_a^d, \quad \text{where} \quad v_a^d = \hat{\hat{\sigma}}_a^d \ x(-|\frac{\hat{\mu}_a^d - \underset{a' \neq a, \, a' \in A}{\max} \hat{\mu}_{a'}^d}{\hat{\hat{\sigma}}_a^d}|) \quad \forall_d \in D$$

and $\boldsymbol{z} = [z^1, \cdots, z^D]^T$ is a reference point. For each objective d, the corresponding reference is the minimum of the current estimated means of all arms minus a small positive value, $\epsilon^d > 0$. The reference z^d for objective d is calculated as:

$$z^d = \underset{1 \leq a \leq |A|}{\min} \hat{\mu}_a^d - \epsilon^d \quad \forall_d \in D$$

The *Cheb*-KG selects the optimal arm a^* that has maximum $f^s(\hat{\mu}_a)$ as:

$$a^*_{Cheb-KG} = \underset{a=1,\cdots,|A|}{\text{argmax}} \ f^s(\hat{\boldsymbol{\mu}}_a) \quad (12)$$

The Algorithm. The pseudocode of the scalarized MOMAB algorithm [5] is given in Fig. 2. Given the type of the scalarized function f, (f is either linear-scalarized-UCB1(LS-UCB1), Chebyshev-scalarized-UCB1 (*Cheb*-UCB1), LS_2-KG, LS_2-KG, or *Cheb*-KG) and the scalarized function set $\boldsymbol{F} = (f^1, \cdots, f^S)$ where each scalarized function f^s has a different weight set, $\boldsymbol{w}^s = (w^{1,s}, \cdots, w^{D,s})$ with $\sum_{d=1}^D w^{d,s} = 1$.

1. **Input:** Horizon of an experiment L;type of scalarized function f;set of scalarized functions $(f^1, \cdots, f^S)$;reward distribution $r_a \sim N(\mu_a, \sigma_a^2)$ $\forall_a \in A$.
2. **Initialize:** For $s = 1$ to S

   plays each arm $Initial$ steps,

   observe: $(r_a)^s$;update: $N^s \leftarrow N^s + 1$; $N_a^s \leftarrow N_a^s + 1$; $(\hat{\mu}_a)^s$; $(\hat{\sigma}_a)^s$

   End
3. **Repeat**
4. **Select a function** s **uniformly at random**
5. **Select the optimal arm** a^* **that maximizes the scalarized function** f^s
6. **Observe: reward vector** r_{a^*}, $r_{a^*} = [r_{a^*}^1, \cdots, r_{a^*}^D]^T$
7. **Update:** $\hat{\mu}_{a^*}$; $\hat{\sigma}_{a^*}$; $N_{a^*}^s \leftarrow N_{a^*}^s + 1$; $N^s \leftarrow N^s + 1$
8. **Compute: unfairness regret; scalarized regret**
9. **Until** L
10. **Output: Unfairness regret; scalarized regret.**

Fig. 2. Algorithm: (Scalarized MOMAB algorithm).

The algorithm in Fig. 2 plays each arm of each scalarized function f^s, *Initial* plays (Step 2). The N^s is the number of times the scalarized function f^s is pulled and N_a^s is the number of times the arm a under the scalarized function f^s is pulled. The $(r_a)^s$ is the reward vector of the pulled arm a which is drawn from a corresponding multivariate normal distribution $N(\mu_a, \sigma_a^2)$ where μ_a is the true mean vector and σ_a^2 is the true variance vector of the reward. The $(\hat{\mu}_a)^s$ and $(\hat{\sigma}_a)^s$ are the estimated mean and standard deviation vectors of the arm a under the scalarized function s, respectively. After initial playing, the algorithm chooses uniformly at random one of the scalarized function (Step 4), selects the optimal arm a^* that maximizes the type of this scalarized function (Step 5) and simulates the selected arm a^*. The estimated mean vector $(\hat{\mu}_{a^*})^s$, estimated standard deviation vector $(\hat{\sigma}_{a^*})^s$, and the number $N_{a^*}^s$ of the selected arm and the number of the pulled scalarized function are updated (Step 7), c.f. Eq. (1) and the unfairness and the scalarized regrets are updated. This procedure is repeated until the end of playing L steps.

5   Experiments

In this section, we experimentally compare Pareto algorithms (Pareto -UCB1 and -KG) and scalarized functions (LS-UCB1, $Cheb$-UCB1, LS_1-KG, LS_2-KG and $Cheb$-KG). At each time step t, the performance measures are: (1) the scalarized and the cumulative scalarized regret (Sect. 2.4) for scalarized functions. (2) the Pareto regret (Sect. 2.4) for Pareto algorithms. (3) the cumulative unfairness regret for scalarization functions. The unfairness regret for Pareto algorithms (Sect. 2.4). The performance measures are the average of M experiments.

We used the algorithm in Fig. 2 for the scalarized functions, and the algorithm in Fig. 1 for the Pareto-KG. To compute the Pareto regret, we need to calculate the virtual distance. The virtual distance dis^* that is added to the mean vector

$\boldsymbol{\mu}_t$ of the pulled arm a_t at time step t (the pulled arm is not element in the Pareto front A^*) in each objective $d = 1, \cdots, D$ can be calculated by firstly ranking all the Euclidean distance dis between the mean vectors of the optimal arms in the Pareto front and zero vector $\mathbf{0} = [0^1, \cdots, 0^D]^T$ as:

$$dis(\boldsymbol{\mu}_1^*, \mathbf{0}) < dis(\boldsymbol{\mu}_2^*, \mathbf{0}) < \cdots < dis(\boldsymbol{\mu}_{|A^*|}^*, \mathbf{0}), \text{ where } dis_1 < dis_2 < \cdots < dis_{|A^*|}$$

Secondly, finding the minimum added distance dis^* which is calculated as:

$$dis^* = dis_1 - dis(\boldsymbol{\mu}_t, \mathbf{0}) \tag{13}$$

where dis_1 is the Euclidean distance between $\mathbf{0}$ vector and the Pareto optimal mean vector $\boldsymbol{\mu}_1^*$, and $dis(\boldsymbol{\mu}_t, \mathbf{0})$ is the Euclidean distance between the mean vector of the pulled arm that is not element in the Pareto front and vector $\mathbf{0}$. Then, add dis^* to the mean vector $\boldsymbol{\mu}_t$ of the pulled arm a_t to create a mean vector that is element in the Pareto front (i.e., $\boldsymbol{\mu}_t^* = \boldsymbol{\mu}_t + \boldsymbol{dis}^*$, where $\boldsymbol{dis}^* = [dis^{*,1}, \cdots, dis^{*,D}]^T$ and $dis^{*,1} = dis^{*,2} = \cdots = dis^{*,D}$) and check if $\boldsymbol{\mu}_t^*$ is a virtual vector that is incomparable with the Pareto front set. If $\boldsymbol{\mu}_t^*$ is incomparable with the mean vectors of Pareto front, then dis^* is the virtual distance, calculate the regret. Otherwise, reduce the added distance to find dis^*, $dis^* = (dis_1 - \frac{dis_2 - dis_1}{1/D}) - dis(\boldsymbol{\mu}_t, \mathbf{0})$, where D is the number of objectives. And, check whether dis^* creates $\boldsymbol{\mu}_t^*$ that is incomparable with the Pareto front. If dis^* does not create a non-dominated arm, then reduce again the dis^* by using dis_3 instead of dis_2 and so on [16].

Shannon entropy measures the unfairness regret, Sect. 2.4. For example, for 2-objective, 6-armed. The true mean vector set is ($\boldsymbol{\mu}_1 = [0.55, 0.5]^T, \boldsymbol{\mu}_2 = [0.53, 0.51]^T, \boldsymbol{\mu}_3 = [0.52, 0.54]^T, \boldsymbol{\mu}_4 = [0.5, 0.57]^T, \boldsymbol{\mu}_5 = [0.51, 0.51]^T, \boldsymbol{\mu}_6 = [0.5, 0.5]^T$). The Pareto front $A^* = \{a_1^*, a_2^*, a_3^*, a_4^*\}$, where a_i^* is an optimal arm. At time step $t = 100$, the optimal number vector $\boldsymbol{N}^*$ of selecting each arm is $\boldsymbol{N}^* = [25, 25, 25, 25, 0, 0]^T$, where each row in the vector $\boldsymbol{N}^*$ gives the optimal number of pulling the corresponding optimal arm. If the number of selecting each arm vector $\boldsymbol{N}$ by an algorithm is $\boldsymbol{N} = [30, 20, 20, 15, 10, 5]^T$ at time step $t = 100$ without playing *initial* steps, then Shannon entropy is 0.0143 [8].

The number of experiments M is 1000. The horizon of each experiment L is 1000. The rewards of each arm $a \in A$ and objective $d \in D$ are drawn from multivariate normal distribution $N(\boldsymbol{\mu}_a, \sigma_a^2)$ where $\boldsymbol{\mu}_a = [\mu_a^1, \cdots, \mu_a^D]^T$ is the unknown true mean vector and $\boldsymbol{\sigma}_a = [\sigma_a^1, \cdots, \sigma_a^D]^T$ is the unknown true standard deviation vector of the reward. The standard deviations of the reward distributions for arms are equal and have small values. For MAB problem, the performance of average cumulative regret of the KG is increased as the standard deviation increases [3], we set the standard deviation σ_a^d of each arm $a \in A$ and objective $d \in D$ to 0.01 when we compare scalarized functions and 0.1 when we compare Pareto algorithms. KG needs the estimated standard deviation $\hat{\sigma}_a$ for each arm a, therefore, each arm is either played initially 2 times, $Initial = 2$ which is the minimum number to estimate the standard deviation or each arm is considered unknown until it is visited $Initial$ times. If the arm a is unknown, the estimated mean of that arm has a maximum value, i.e. $\hat{\mu}_a^d = \max_{d \in D} \mu_a^d, \forall_a \in A$ and the estimated standard deviation, i.e. $\hat{\sigma}_a^d = \max_{d \in D} \sigma_a^d, \forall_a \in A$ to increase the exploration of arms. We compare the different setting for KG and found out that playing each arm initially 2 times, the scalarized and Pareto regret performances of KG are increased, therefore, we used this to compare with UCB1. For UCB1 (for Pareto-UCB1, LS-UCB1 and $Cheb$-UCB1) each arm is played initially two times, i.e. $Initial = 2$ to get fair comparison with KG policy. After playing each arm two times, the estimated mean of arms are calculated and the scalarized or Pareto regret is computed. The number of Pareto optimal arms $|A^*|$ is unknown to the agent, therefore, $|A^*| = |A|$. We consider 11 weight sets which are uniformly random spread sampling in the weighted space [14] for the scalarization functions as [16]. For instance, for number of objectives equals 2, the total weight set can be set to $W = \{(1,0)^T, (0.9,0.1)^T, \cdots, (0,1)^T\}$. For $Cheb$-UCB1 and $Cheb$-KG, the parameter $\epsilon \in [0,0.1]$ for all the experiments as [5].

5.1   Non-Convex Mean Vector Set

Experiment 1. With number of objectives D equals 3, number of arms $|A|$ equals 50, number of optimal arms $|A^*|$ equals 20 and a non-convex Pareto mean set.

First, we compare the scalarized functions (LS-UCB1, $Cheb$-UCB1, LS_1-KG, LS_2-KG and $Cheb$-KG). Figure 3 gives the scalarized and the cumulative scalarized regret performance using equal standard deviation of the reward distribution σ_a^d for each arm $a \in A$ and each objective $d \in D$, i.e. $\sigma_a^d = 0.01$. At each time step t, the scalarized regret and the cumulative scalarized regret are the average of 1000 experiments. The y-axis is either the scalarized regret performance or the cumulative scalarized regret performance. The x-axis is the horizon of experiments. Figure 3 shows: (1) according to the scalarized regret, $Cheb$-KG is the best scalarized function in the long run. Although, $Cheb$-KG performs as same as $Cheb$-UCB1 according to the cumulative scalarized regret. (2) LS-UCB1 is the worst scalarized function. (3) LS_1-KG outperforms LS_2-KG where the scalarized and the cumulative regrets are decreased. (4) Although, the ϵ parameter of the Chebyshev scalarized-KG and -UCB1 is set to a fixed value, Chebyshev scalarization functions ($Cheb$-UCB1 and $Cheb$-KG) perform better than the linear scalarization functions (LS-UCB1, LS_1-KG and LS_2-KG).

Figure 4 gives the scalarized unfairness and the cumulative scalarized unfairness regret performance using the standard deviation $\sigma_a^d = 0.01$ for each arm a and objective d. At each time step t, the unfairness scalarized regret and the cumulative unfairness scalarized regret are the average of 1000 experiments. The y-axis is either the unfairness scalarized regret performance or the cumulative unfairness scalarized regret performance. The x-axis is the horizon of experiments. Figure 4 shows: (1) $Cheb$-UCB1 is the worst scalarized function, while $Cheb$-KG and LS_2-KG are the best scalarized functions. (2) LS-UCB1 outperforms $Cheb$-UCB1 where the cumulative unfairness regret is decreased. (3) $Cheb$-KG performs as same as LS_2-KG and they outperform LS_1-KG according to the cumulative unfairness regret performance. According to the unfairness regret performance, scalarized KG functions perform evenly in the long run.

Second, we compare the Pareto algorithms (Pareto-UCB1 and Pareto-KG). Figure 5 gives the Pareto and the unfairness regret performance using equal

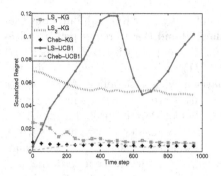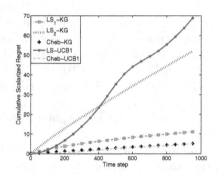

Fig. 3. Performance comparison on 3-objective, 50-armed with a non-convex mean vector set. Left figure shows the scalarized regret performance. Right figure shows the cumulative scalarized regret performance.

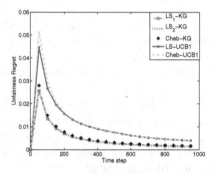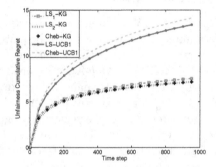

Fig. 4. Performance comparison on 3-objective, 50-armed with a non-convex mean vector set. Left sub-figure shows the unfairness regret performance of the scalarized functions. Right sub-figure shows the cumulative unfairness regret performance of the scalarized functions.

standard deviation of the reward distribution σ_a^d for each arm $a \in A$ and each objective $d \in D$, i.e. $\sigma_a^d = 0.1$. The y-axis is either the Pareto regret or the unfairness regret performance. The x-axis is the horizon of experiments. At each time step t, the Pareto regret and the unfairness regret are the average of 1000 experiments. Figure 5 shows Pareto-KG outperforms Pareto-UCB1 according to both the Pareto and unfairness regret performances. Note that, according to the cumulative Pareto regret and the cumulative unfairness regret, Pareto-KG outperforms Pareto-UCB1. Since, the cumulative Pareto and unfairness regret at time step t are the cumulative summation of the Pareto and the unfairness regrets from time step equals 1 till time step equals t, respectively.

5.2   Convex Mean Vector Set

Experiment 2. With number of objectives $D = 3$, number of arms $|A| = 50$, number of optimal arms $|A^*| = 20$ and a convex Pareto mean set.

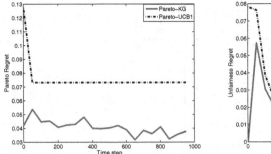
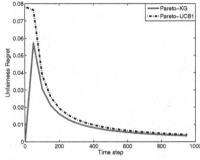

Fig. 5. Performance comparison on 3-objective, 50-armed with a non-convex mean vector set. Left figure shows the Pareto regret performance. Right figure shows the unfairness regret performance.

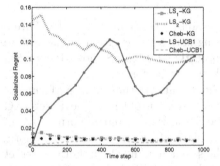
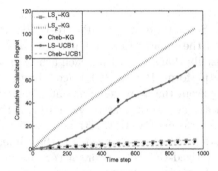

Fig. 6. Performance comparison on 3-objective, 50-armed with a convex mean vector set. Left figure shows the scalarized regret performance. Right figure shows the cumulative scalarized regret performance.

First, we compare the scalarized functions. Figure 6 gives the scalarized and the cumulative scalarized regret performance using the standard deviation of the reward distribution $\sigma_a^d = 0.01$ for each arm a and objective d. Figure 6 shows: (1) according to the cumulative scalarized regret, $Cheb$-UCB1 is the best scalarized function. Although, $Cheb$-UCB1 performs as same as $Cheb$-KG and LS_1-KG almost after time step $t = 300$ according to the scalarized regret performance. (2) LS_2-KG is the worst scalarized function. (3) LS-UCB1 outperforms LS_2-KG.

Figure 7 gives the scalarized and the cumulative scalarized unfairness regret performance using the standard deviation $\sigma_a^d = 0.01$ for each arm a and objective d. Figure 7 shows: (1) according to the cumulative unfairness regret performance, LS_1-KG is the best scalarized function and $Cheb$-UCB1 is the worst one. (2) after almost time step $t = 600$, LS_1-KG performs as same as $Cheb$-KG and LS_2-KG, according to the unfairness regret performance. (3) scalarized-KG functions perform better than the scalarized-UCB1, the unfairness and the cumulative unfairness regrets are decreased dramatically.

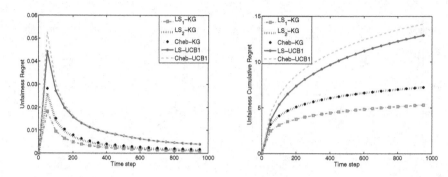

Fig. 7. Performance comparison on 3-objective, 50-armed with a convex mean vector set. Left figure shows the unfairness regret performance and right figure shows the cumulative unfairness regret performance of the scalarized functions.

Second, we compare Pareto-UCB1 and Pareto-KG. Figure 8 gives the Pareto and the unfairness regret performance using standard deviation of the reward distribution $\sigma_a^d = 0.1$ for each arm a and objective d. Figure 8 shows Pareto-KG outperforms Pareto-UCB1 according to the regret performances.

From the above experiments, we conclude that: (1) the regret performance measures are decreased using KG in the MOMAB problems. The Pareto-KG outperforms Pareto-UCB1 according to both the Pareto and unfairness regret performances, and most of the time, scalarized-KG functions outperform scalarized-UCB1 functions according to the scalarized and the unfairness regret performances. (2) according to the scalarized and cumulative scalarized regrets, scalarized $Cheb$-UCB1 function outperforms LS-UCB1 function. According to the unfairness and cumulative unfairness regrets, scalarized LS-UCB1 function outperforms $Cheb$-UCB1 function. (3) according to the scalarized and cumulative scalarized regrets, scalarized $Cheb$-KG and LS_1-KG functions perform better than the LS_2-KG function. According to the unfairness and cumulative unfairness regrets, the performance of LS_1-KG function depends on the used mean vectors set (i.e., either non-convex or convex). With a non-convex mean vector set, LS_1-KG performs almost as same as the $Cheb$-KG and LS_2-KG according to the unfairness regret. With a convex mean vector set, LS_1-KG performs better than the $Cheb$-KG and LS_2-KG according to the unfairness regret performance. (4) the regret performance measures of the KG in the MOMAB is increased by using a non-convex mean vector set. While, the regret performance measures of the UCB1 in the MOMAB is increased by using a convex mean vector set.

The intuition of being KG outperforms UCB1 is the exploration bound. For an arm a, UCB1 computes an equal exploration bound for each objective $d \in D$. The exploration bound of UCB1 depends on the time step t and the number of times N_a arm a is pulled and it will be high if the arm a is less selected. In contrast, the added exploration bound of an arm a to its estimated mean vector in each objective $d \in D$ by KG will be not the same value in each objective, i.e.

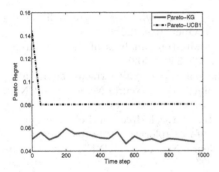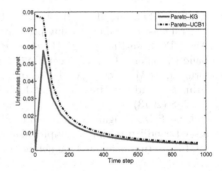

Fig. 8. Performance comparison on 3-objective, 50-armed with a convex mean vector set. Left figure shows the Pareto regret performance. Right figure shows the unfairness regret performance.

each objective has different exploration bound. The exploration bound of KG policy for arm a depends on the estimated mean of all other arms and on the estimated variance of the arm a. The exploration bound is large if the variance of arm a is low, or if the estimated mean of arm a exceeds in the future. Thus, KG trades off efficiently between exploration and exploitation.

6 Conclusions and Future Work

We have introduced the MOMAB problem for the Knowledge Gradient (KG) policy to fine tune the trade off between identifying (exploration) and selecting fairly (exploitation) the Pareto front. We have presented the regret measures in the MOMAB problem, Pareto dominance relations and scalarized functions. We presented the multi-objective UCB1 policy to experimentally compare the two policies. We proposed Pareto-KG that uses the Pareto dominance relation to partially order the arms. We proposed two types of linear scalarized-KG (linear scalarized-KG across arms (LS_1-KG) and linear scalarized-KG across dimensions (LS_2-KG)) and Chebyshev-scalarized-KG ($Cheb$-KG). Finally, we experimentally compared Pareto and scalarized KG and with homologues UCB1 policies and concluded that the average regret is improved using KG policy in the MOMAB problem. Future work must provide theoretical analysis for the KG in the MOMAB and must compare UCB1 and KG policy on stochastic correlated MOMAB problems.

References

1. Sutton, R.S., Barto, A.G.: Reinforcement Learning: An Introduction. MIT Press, Cambridge (1998)
2. Yahyaa, S.Q., Manderick, B.: The exploration vs exploitation trade-off in the multi-armed bandit problem: an empirical study. In: 20th European Symposium on Artificial Neural Networks (ESANN), pp. 549–554 (2012)

3. Ryzhov, I.O., Powell, W.B., Frazier, P.I.: The knowledge-gradient policy for a general class of online learning problems. J. Oper. Res. (2011)
4. Auer, P., Cesa-Bianchi, N., Fischer, P.: Finite-time analysis of the multiarmed bandit problem. J. Mach. Learn. **47**(2–3), 235–256 (2002)
5. Drugan, M.M., Nowe, A.: Designing multi-objective multi-armed bandits algorithms: a study. In: International Joint Conference on Neural Networks (IJCNN), pp. 1–8 (2013)
6. Yahyaa, S.Q., Drugan, M.M., Manderick, B.: The scalarized multi-objective multi-armed bandit problem: an empirical study of its exploration vs. exploration trade-off. In: International Joint Conference on Neural Networks (IJCNN), pp. 2290–2297 (2014)
7. Yahyaa, S.Q., Drugan, M.M., Manderick, B.: Linear scalarized knowledge gradient in the multi-objective multi-armed bandits problem. In: 22th European Symposium on Artificial Neural Networks (ESANN) (2014)
8. Yahyaa, S.Q., Drugan, M.M., Manderick, B.: Annealing-pareto multi-objective multi-armed bandits algorithm. In: IEEE Symposium on Adaptive Dynamic Programming and Reinforcement Learning (ADPRL), pp. 1–8 (2014)
9. Zitzler, E., Thiele, L., Laumanns, M., Fonseca, C.M., Da Fonseca, V.G.: Performance assessment of multiobjective optimizers: an analysis and review. IEEE Trans. Evol. Comput. **7**, 117–132 (2002)
10. Eichfelder, G.: Adaptive Scalarization Methods in Multiobjective Optimization. Springer, Heidelberg (2008)
11. Powell, W.B., Ryzhov, I.O.: Optimal Learning. Willey, Canada (2012)
12. Kuleshov, V., Precup, D.: Algorithms for multi-armed bandit problems. J. (2014). CoRR abs/1402.6028
13. Powell, W.B.: Approximate Dynamic Programming: Solving the Curses of Dimensionality. Wiley, New York (2007)
14. Das, I., Dennis, J.E.: A closer look at drawbacks of minimizing weighted sums of objectives for pareto set generation in multicriteria optimization problems. Struct. Optim. **14**(1), 63–69 (1997)
15. Miettinen, K.: Nonlinear Multiobjective Optimization. International Series in Operations Research and Management Science. Springer, Heidelberg (1999)
16. Yahyaa, S.Q., Drugan, M.M., Manderick, B.: Knowledge gradient for multi-objective multi-armed bandit algorithms. In: 6th International Conference on Agents and Artificial Intelligence (ICAART) (2014)

Extensibility Based Multiagent Planner with Plan Diversity Metrics

Jan Tožička[(✉)], Jan Jakubův, Karel Durkota, and Antonín Komenda

Agent Technology Center, Czech Technical University, Prague, Czech Republic
{jan.tozicka,jan.jakubuv,karel.durkota,
antonin.komenda}@agents.fel.cvut.cz

Abstract. Coordinated sequential decision making of a team of cooperative agents is described by principles of multiagent planning. In this work, we extend the MA-STRIPS formalism with the notion of *extensibility* and reuse a well-known initiator–participants scheme for agent negotiation. A multiagent extension of the Generate-And-Test principle is used to distributively search for a coordinated multiagent plan. The generate part uses a novel plan quality estimation technique based on metrics often used in the field of diverse planning. The test part builds upon planning with landmark actions by compilation to classic planning. We designed a new multiagent planning domain which illustrates the basic properties of the proposed multiagent planning approach. Finally, our approach was experimentally evaluated on four classic IPC benchmark domains modified for multiagent settings. The results show (1) which combination of plan quality estimation and (2) which diversity metrics provide the best planning efficiency.

Keywords: Multiagent planning · Diverse planning · Planning landmarks

1 Introduction

Multiagent planning is a specific form of distributed planning and problem solving, which was summarized by [6]. Multiagent planning research and literature focused mostly on the *coordination* part of the problem while the *synthesis* part dealing with a particular ordering of actions was studied in the area of classic planning.

The coordination was, for instance, studied in the well-known General Partial Global Planning in [4] and with additional domain-specific information as TALPlanner in [5]. Brafman and Domshlak in [3] proposed the first fusion of the coordination and synthesis parts for domain-independent multiagent planning with deterministic actions. The approach was based on the classic planning formalism STRIPS [7] extended to multiagent settings denoted as MA-STRIPS. [3] also proposed a solution for the coordination part of the problem by translation to a Distributed Constraint Satisfaction Problem (DCSP). Since the paper was

© Springer-Verlag Berlin Heidelberg 2015
N.T. Nguyen et al. (Eds.): TCCI XX, LNCS 9420, pp. 117–139, 2015.
DOI: 10.1007/978-3-319-27543-7_6

focused primarily on theoretical analysis of computational complexity of MA-STRIPS problems, several algorithmic approaches appeared later in other papers, e.g., in [10] or [14].

In this paper, we propose a novel algorithmic approach to multiagent planning for problems described in MA-STRIPS based on the principle of classic multiagent negotiation protocols such as Contract Net with one agent playing a role of an initiator and the rest acting as participants. The approach can be seen as a protocol describing distribution of the *Generate-And-Test Search* which was a base principle also in multiagent planners described by [10] (using DCSP for the coordination part and classic planner for the generation part) and by [11] (using backtracking search for the coordination part and planning graphs for the generation part).

The contribution of our work is in the way of how plan candidates are generated and tested. Our generative process uses estimation of quality of generated plans based on metrics of diverse planning (particularly from [2,13]). In other words, the idea is to generate good-quality plans and avoid low-quality ones. The quality measure is based on the history of answers of the participants who were trying to extend the initial plan. Therefore it can be understood as a learning of the initiator agent to generate plan candidates which can be more likely extended by more participant agents to a final solution.

The testing part utilizes planning with landmarks (similarly as used by [10]). The difference is that we translate a planning problem with landmarks into an ordinary planning problem, which can be then solved by a classic planner. Usually, landmarks are incorporated into planners as special heuristics as in [12]. However, our translation enables a straightforward incorporation of externally defined landmarks, which is required by the proposed planning protocol.

Finally, we provide experimental evaluation of the planner on a newly designed planning domain *tools* and four planning domains from International Planning Competition extended for multiagent planning.

2 Multiagent Planning

We consider a number of *cooperative* and *coordinated* agents featuring distinct sets of capabilities (actions) which concurrently plan and execute their local plans in order to achieve a joint goal. The environment wherein the agents act is *classic* with *deterministic* actions. The following formal preliminaries restate the MA-STRIPS problem [3] and define *local planning problems* and plan *extensibility* required for the following sections.

2.1 Planning Problem

An MA-STRIPS *planning problem* Π is a quadruple $\Pi = \langle P, \{\alpha_i\}_{i=1}^n, I, G \rangle$, where P is a set of facts, α_i is the set of actions of i-th agent, $I \subseteq P$ is an initial state, and $G \subseteq P$ is a set of goal facts.

An *action* an agent can perform is a quadruple containing a unique action id and three subsets of P which in turn denote the set of *preconditions*, the set of *add effects*, and the set of *delete effects*. Action ids are arbitrary atomic objects and we always consider ids to be unique within a given problem. Selector functions $\mathsf{id}(a), \mathsf{pre}(a), \mathsf{add}(a)$, and $\mathsf{del}(a)$ are defined so that the following holds.

$$a = \langle \mathsf{id}(a), \mathsf{pre}(a), \mathsf{add}(a), \mathsf{del}(a) \rangle$$

Moreover let $\mathsf{eff}(a) = \mathsf{add}(a) \cup \mathsf{del}(a)$.

An *agent* is identified with its capabilities, in other words, the i-th agent $\alpha_i = \{a_1, \ldots, a_m\}$ is characterized by a finite repertoire of actions it can preform in the environment. We use metavariable α to range over agents from Π. A *planning state* s is a finite set of facts and we say that fact p *holds in* s, or that p *is valid in* s, iff $p \in s$. When $\mathsf{pre}(a) \subseteq s$ then *state progression* function γ is defined classically as $\gamma(s, a) = (s \setminus \mathsf{del}(a)) \cup \mathsf{add}(a)$.

Example 1. We shall demonstrate definitions of this section on a simple logistic problem involving three locations **prague**, **brno**, and **ostrava**, two vehicles **plane** and **truck**, and one **crown** to be delivered from **prague** to **ostrava**. A **plane** can travel from **prague** to **brno** and back, while a **truck** provides connection between **brno** and **ostrava**. Instances of actions $\mathtt{fly}(loc_1, loc_2)$ and $\mathtt{drive}(loc_1, loc_2)$ describe movement of **plane** and **truck** respectively, while $\mathtt{load}(veh, loc)$ and $\mathtt{unload}(veh, loc)$ describe loading and unloading of **crown** by a given vehicle at a given location.

We define two agents *Plane* and *Truck*. The agents are defined by sets of executable actions as follows.

Plane = {	*Truck* = {
fly(prague, brno),	drive(brno, ostrava),
fly(brno, prague),	drive(ostrava, brno),
load(plane, prague),	load(truck, brno),
load(plane, brno),	load(truck, ostrava),
unload(plane, prague),	unload(truck, brno),
unload(plane, brno)	unload(truck, ostrava)
}	}

Aforementioned actions are defined using facts $\mathtt{at}(veh, loc)$ to describe possible vehicle locations, and facts $\mathtt{in}(crown, loc)$ and $\mathtt{in}(crown, veh)$ to describe positions of **crown**. We omit action ids in examples when no confusion can arise. For example, we have the following.

$$\mathtt{fly}(loc_1, loc_2) = \langle \qquad\qquad \mathtt{load}(veh, loc) = \langle$$
$$\{\mathtt{at}(plane, loc_1)\}, \qquad\qquad \{\mathtt{at}(veh, loc), \mathtt{in}(crown, loc)\},$$
$$\{\mathtt{at}(plane, loc_2)\}, \qquad\qquad \{\mathtt{in}(crown, veh)\},$$
$$\{\mathtt{at}(plane, loc_1)\} \qquad\qquad \{\mathtt{in}(crown, loc)\}$$
$$\rangle \qquad\qquad\qquad\qquad \rangle$$

The initial state and the goal are given as follows.

$$I = \{\text{at}(\text{plane}, \text{prague}), \text{at}(\text{truck}, \text{brno}), \text{in}(\text{crown}, \text{prague})\}$$
$$G = \{\text{in}(\text{crown}, \text{ostrava})\} \qquad\qquad \boxtimes$$

2.2   Public and Internal Classification

In MA-STRIPS multiagent planning, each fact is classified either as *public* or as *internal* out of computational or privacy concerns. MA-STRIPS specifies this classification as follows. A fact is *public* when it is mentioned by actions of at least two different agents. A fact is *internal for agent* α when it is not public but mentioned by some action of α. A fact is *relevant for* α when it is either public or internal for α. Relevant facts contain all the facts which agent α needs to understand, because other facts are internal for other agents and thus do not directly concern α. The following formally defines public, internal, and relevant facts.

Definition 1. *Let* $\Pi = \langle P, \{\alpha_i\}_{i=1}^n, I, G \rangle$. *The sets of facts* facts(a) *and* facts(α) *are defined for every action* a *and agent* α *from* Π *as follows.*

$$\text{facts}(a) = \text{pre}(a) \cup \text{add}(a) \cup \text{del}(a)$$
$$\text{facts}(\alpha) = \bigcup_{a \in \alpha} \text{facts}(a)$$

Furthermore, the set pub-facts *of public facts, and sets* int-facts(α) *and* rel-facts(α) *of facts internal for agent* α *and relevant for* α *are defined as follows.*

$$\text{pub-facts} = \bigcup_{i \neq j}(\text{facts}(\alpha_i) \cap \text{facts}(\alpha_j)) \qquad (i, j \in \{1, \dots, n\})$$
$$\text{int-facts}(\alpha) = \text{facts}(\alpha) \setminus \text{pub-facts}$$
$$\text{rel-facts}(\alpha) = \text{int-facts}(\alpha) \cup \text{pub-facts} \qquad (= \text{facts}(\alpha) \cup \text{pub-facts})$$

$\boxtimes$

It is common in literature [9] to require that all the goals are public. An MA-STRIPS problem with internal goals can be easily transformed to an equivalent problem without internal goals and thus we omit internal goals in formal presentation. Hence pub-facts is defined as the minimal superset of the intersection from the definition that satisfies $G \subseteq$ pub-facts. In the rest of this paper we assume $G \subseteq$ pub-facts and also another simplification common in literature [3] which says that α_i are pairwise disjoint[1].

Example 2. In our running example, the only fact shared by the two agents is in(crown, brno). As we require $G \subseteq$ pub-facts we have the following facts classification.

$$\text{pub-facts} = \{\text{in}(\text{crown}, \text{brno}), \text{in}(\text{crown}, \text{ostrava})\}$$
$$\text{int-facts}(Plane) = \{\text{at}(\text{plane}, \text{prague}), \text{at}(\text{plane}, \text{brno}),$$
$$\text{in}(\text{crown}, \text{prague}), \text{in}(\text{crown}, \text{plane})\} \qquad \boxtimes$$

---

[1] This rules out *joint actions*. Any MA-STRIPS problem with joint actions can be translated to an equivalent problem without joint actions. However, a solution that would take advantage joint actions is left for future research.

MA-STRIPS further extends this classification of facts to actions as follows. An action is *public* when it has a public effect, otherwise it is *internal*. Strictly speaking, MA-STRIPS defines an action as public whenever it mentions a public fact even in a precondition (that is, when $\mathsf{facts}(a) \cap \mathsf{pub\text{-}facts} \neq \emptyset$). However, as the method of multiagent planning with external actions presented in this work does not rely on synchronization on public preconditions, we can allow actions with only public preconditions as internal. For our planner it is sufficient to know that internal actions do not *modify* public state. The following formally defines public and internal actions.

Definition 2. *Let* $\Pi = \langle P, \{\alpha_i\}_{i=1}^n, I, G \rangle$ *and* $A = \bigcup_{i=1}^n \alpha_i$. *The set* pub-actions *of public actions and the set* int-actions(α) *of actions internal for* α *are defined as follows.*

$$\mathsf{pub\text{-}actions} = \{a \in A : \mathsf{eff}(a) \cap \mathsf{pub\text{-}facts} \neq \emptyset\}$$
$$\mathsf{int\text{-}actions}(\alpha) = \alpha \setminus \mathsf{pub\text{-}actions}$$

⊠

2.3 Local Planning Problems

In MA-STRIPS problems, agent actions are assumed to manipulate a shared global state when executed. In this section, we construct a *local planning problem* for every agent α. Each local planning problem for α is a classic STRIPS problem where agent α has its own internal copy of the global state and where each agent is equipped with *external actions* which provide information about public actions of other agents. These local planning problems allow us to divide an MA-STRIPS problem to several STRIPS problems which can be solved separately by a classic planner. In this paper, we describe a way how to find a solution of an MA-STRIPS problem but it does not address the question of *execution* of a plan in some real-world environment.

The *projection* $F \triangleright \alpha$ of set of facts F to agent α is the restriction of F to the facts relevant for α. Hence projection removes from F facts not relevant for α and thus it represents F as understood by agent α. The *projection* $a \triangleright \alpha$ of action a to agent α removes from a facts not relevant for α, again representing a as seen by α. The projections are formally defined as follows.

Definition 3. *Let* $\Pi = \langle P, \{\alpha_i\}_{i=1}^n, I, G \rangle$. *Let* F *be an arbitrary set* $F \subseteq P$ *of facts and let* a *be an action from* Π. *The projection* $F \triangleright \alpha$ *of* F *to agent* α, *and the* projection $a \triangleright \alpha$ *of action* a *to* α *are defined as follows.*

$$F \triangleright \alpha = F \cap \mathsf{rel\text{-}facts}(\alpha) \qquad a \triangleright \alpha = \langle \mathsf{id}(a), \mathsf{pre}(a) \triangleright \alpha, \ \mathsf{add}(a) \triangleright \alpha, \ \mathsf{del}(a) \triangleright \alpha \rangle$$

⊠

Note that $a \triangleright \alpha = a$ when $a \in \alpha$. Hence projection to α alters only actions of agents other than α. Also note that action ids are preserved under projection.

Example 3. In our example we have the following.

$$\mathsf{fly}(\mathtt{prague}, \mathtt{brno}) \triangleright Plane = \mathsf{fly}(\mathtt{prague}, \mathtt{brno})$$
$$\mathsf{fly}(\mathtt{prague}, \mathtt{brno}) \triangleright Truck = \langle \emptyset, \emptyset, \emptyset \rangle$$
$$\mathsf{load}(\mathtt{truck}, \mathtt{brno}) \triangleright Plane = \langle \{\mathsf{in}(\mathtt{crown}, \mathtt{brno})\}, \emptyset, \{\mathsf{in}(\mathtt{crown}, \mathtt{brno})\} \rangle$$
$$\mathsf{unload}(\mathtt{truck}, \mathtt{ostrava}) \triangleright Plane = \langle \emptyset, \{\mathsf{in}(\mathtt{crown}, \mathtt{ostrava})\}, \emptyset \rangle \qquad ⊠$$

In *multiagent planning with external actions*, every agent α is from the beginning equipped with projections of other agents public actions. These projections, which we call *external actions*, describe how agent α sees effects of public actions of other agents. In a local planning problem, an agent needs external actions so that he can create a plan which contains also actions of other agents. The set of actions in a local planning problem of agent α simply contains actions of agent α together with external actions of α. The actions in a local planning problem of agent α are called *actions relevant for* α. The above action sets are formally defined as follows.

Definition 4. *Let* $\Pi = \langle P, \{\alpha_i\}_{i=1}^n, I, G \rangle$. *The set* ext-actions($\alpha$) *of external actions of agent* α, *and the set* rel-actions(α) *of relevant actions of* α *are defined as follows.*

$$\text{ext-actions}(\alpha) = \{a \triangleright \alpha : a \in \text{pub-actions } and \ a \notin \alpha\}$$
$$\text{rel-actions}(\alpha) = \alpha \cup \text{ext-actions}(\alpha) \hspace{2cm} \boxtimes$$

Example 4. In our example, all the actions arranging vehicle movements are internal. Public are only the actions providing package treatment at public locations (brno, ostrava). Hence the set of public actions of agent *Plane* (that is, the set *Plane* $\cap$ pub-actions) is as follows.

$$\{ \text{load(plane, brno), unload(plane, brno)} \}$$

In contrast, agent *Truck* has the following four public actions.

$$\{ \text{load(truck, brno),} \quad \text{unload(truck, brno),}$$
$$\text{load(truck, ostrava), unload(truck, ostrava)} \}$$

Hence rel-actions(*Plane*) has 10 actions (six own plus four external) while rel-actions(*Truck*) has only 8 actions (six own plus two external). $\hspace{1cm} \boxtimes$

Now it is easy to define a *local planning problem* $\Pi \triangleright \alpha$ of agent α also called *projection of* Π *to* α. The set of facts P and the initial state I are restricted to those facts relevant for α. There is no need to restrict the goal G because all the goal facts are public and thus relevant for all the agents.

Definition 5. *The* projection *of* Π *to* α *is defined as follows.*

$$\Pi \triangleright \alpha = \langle P \triangleright \alpha, \ \text{rel-actions}(\alpha), \ I \triangleright \alpha, \ G \rangle \hspace{2cm} \boxtimes$$

Example 5. In our example we have

$$I \triangleright Plane = \{\text{at(plane, prague), in(crown, prague)}\}$$
$$\Pi \triangleright Plane = \langle P \triangleright Plane, \text{rel-actions}(Plane), I \triangleright Plane, G \rangle$$

Projection $\Pi \triangleright$ *Truck* is defined similarly. $\hspace{4cm} \boxtimes$

2.4   Planning with External Actions

We would like to solve agent local problems separately and compose local solutions to a global solution of Π. However, not all local solutions can be easily composed to a solution of Π. Concepts of *public plans* and their *extensibility* help us to recognize local solutions which are conductive to this aim.

A *plan* π is a sequence of actions $\langle a_1, \ldots, a_k \rangle$. A plan π defines an order in which the actions are executed by their unique owner agents. It is assumed that independent actions can be executed in parallel. A *solution* of Π is a plan π whose execution transforms the initial state I to the state s such that $G \subseteq s$. A *local solution* of agent α is a solution of the local planning problem $\Pi \triangleright \alpha$. Let $\mathsf{sols}(\Pi)$ and $\mathsf{sols}(\Pi \triangleright \alpha)$ denote the sets of all the solutions of MA-STRIPS problem Π and all the local solutions of α respectively.

Example 6. Let us consider the following plans.

$\pi_0 = \langle$ load(plane, prague), fly(prague, brno), unload(plane, brno),
      load(truck, brno), drive(brno, ostrava), unload(truck, ostrava) $\rangle$
$\pi_1 = \langle$ unload(truck, ostrava) $\triangleright$ *Plane* $\rangle$
$\pi_2 = \langle$ unload(plane, brno) $\triangleright$ *Truck*, load(truck, brno),
      drive(brno, ostrava), unload(truck, ostrava) $\rangle$

It is easy to check that π_0 is a solution of our example MA-STRIPS problem Π. Plan π_1 is a solution of projection $\Pi \triangleright$ *Plane* because *Plane*'s projection unload(truck, ostrava) $\triangleright$ *Plane* of *Truck*'s public action simply produces the goal state out of the blue. Finally, $\pi_2 \in \mathsf{sols}(\Pi \triangleright$ *Truck*).   ☒

A *public plan* σ is a plan that contains only public actions. A public plan can be seen as a solution outline that captures execution order of public actions while ignoring agents internal actions. A public plan can be safely sent to any agent because it contains only public information. In order to avoid confusions between public and external versions of the same action, we formally define public plans to contain only public action *ids*. Hence a public plan can be interpreted differently in the context of MA-STRIPS problem Π, where each *id* is interpreted directly as an action from Π, and in the context of a local problem $\Pi \triangleright \alpha$ where an *id* corresponds to an action from $\Pi \triangleright \alpha$ (possibly projected action). From a plan π of Π (or a plan of $\Pi \triangleright \alpha$) we can define the *public projection* $\pi \triangleright \star$ of π as the sequence of all public action *ids* from π preserving their order. Public projection of a plan thus removes any internal actions from π. A formal definition follows.

Definition 6. *A public plan σ is a sequence of public action* ids. *Given a plan π of Π (or of $\Pi \triangleright \alpha$), the public projection $\pi \triangleright \star$ of π is defined to be the following public plan.*

$$\pi \triangleright \star = \langle \mathsf{id}(a) : a \in \pi \ and \ a \in \mathsf{pub\text{-}actions} \rangle \qquad ☒$$

Example 7. In our example we know that $\pi_0 \in \mathsf{sols}(\Pi)$ and $\pi_1 \in \mathsf{sols}(\Pi \rhd \textit{Plane})$ and $\pi_2 \in \mathsf{sols}(\Pi \rhd \textit{Truck})$. Thus we can construct the following public solutions.

$$\pi_0 \rhd \star = \langle \mathsf{id}(\mathsf{unload}(\mathsf{plane}, \mathsf{brno})),$$
$$\mathsf{id}(\mathsf{load}(\mathsf{truck}, \mathsf{brno})),$$
$$\mathsf{id}(\mathsf{unload}(\mathsf{truck}, \mathsf{ostrava}))\rangle$$
$$\pi_1 \rhd \star = \langle \mathsf{id}(\mathsf{unload}(\mathsf{truck}, \mathsf{ostrava}))\rangle$$
$$\pi_2 \rhd \star = \langle \mathsf{id}(\mathsf{unload}(\mathsf{plane}, \mathsf{brno})),$$
$$\mathsf{id}(\mathsf{load}(\mathsf{truck}, \mathsf{brno})),$$
$$\mathsf{id}(\mathsf{unload}(\mathsf{truck}, \mathsf{ostrava}))\rangle$$

Note that $\pi_0 \rhd \star = \pi_2 \rhd \star$ and also note that we have omitted the projection operator ($\rhd$) because ids are preserved under projection. ☒

From every solution π of Π (or $\Pi \rhd \alpha$) we can construct a uniquely determined public plan $\sigma = \pi \rhd \star$. In contrast, for a single public plan σ there might be more than one, or none, solutions π with public projection σ. A public plan σ is called *internally extensible* when there is a solution of Π with public projection σ. Similarly, when there is a solution of $\Pi \rhd \alpha$ with public projection σ, then σ is called *internally α-extensible*. Extensible public plans give us an order of public actions which is acceptable for all the agents. Thus internally extensible public plans are very close to solutions of Π and it is relatively easy to construct a solution of Π once we have an internally extensible public plan. Hence our algorithms will aim at finding internally extensible public plans. The following formally defines public plan extensibility.

Definition 7. *Let σ be a public plan of* Π.

$$\sigma \textit{ is internally extensible} \quad \textit{iff} \;\; \exists \pi \in \mathsf{sols}(\Pi) : \pi \rhd \star = \sigma$$
$$\sigma \textit{ is internally } \alpha\textit{-extensible} \quad \textit{iff} \;\; \exists \pi \in \mathsf{sols}(\Pi \rhd \alpha) : \pi \rhd \star = \sigma \qquad ☒$$

Example 8. In our example we can see that $\pi_0 \rhd \star$ is extensible because it was constructed from the solution of Π. For the same reason we see that $\pi_1 \rhd \star$ is *Plane*-extensible and $\pi_2 \rhd \star$ is *Truck*-extensible. It is easy to see that $\pi_2 \rhd \star$ is also *Plane*-extensible. However, $\pi_1 \rhd \star$ is not *Truck*-extensible because *Truck* needs to execute other public actions prior to $\mathsf{unload}(\mathsf{truck}, \mathsf{ostrava})$. ☒

The following proposition states the correctness of the multiagent planning with external actions. It establishes the relationship between (internally) extensible and α-extensible plans. Its direct consequence is that to find a solution of Π it is enough to find a local solution $\pi_\alpha \in \mathsf{sols}(\Pi \rhd \alpha)$ which is β-extensible for every other agent β. A constructive proof follows.

Theorem 1 ([15]). *Public plan σ of Π is internally extensible if and only if σ is internally α-extensible for every agent α.*

Proof ($\Rightarrow$). *When σ is internally extensible then there is $\pi \in \mathsf{sols}(\Pi)$ such that $\pi \rhd \star = \sigma$. Let α be arbitrary but fixed. Let us construct plan π_α of $\Pi \rhd \alpha$ from π*

by removing internal actions of agents other than α, *and by applying projection to the remaining actions as follows.*

$$\pi_\alpha = \langle a \triangleright \alpha : a \in \pi \text{ and } a \in \text{pub-actions} \cup \text{int-actions}(\alpha)\rangle$$

Clearly $\pi_\alpha \triangleright \star = \sigma$ *because* π_α *preserves the order of public actions. To prove* $\pi_\alpha \in \text{sols}(\Pi \triangleright \alpha)$ *we first observe that no action* b *internal for* $\beta \neq \alpha$ *can change state* s *of* Π *in a way observable by* α, *that is,* $\gamma(s, b) \triangleright \alpha = s \triangleright \alpha$. *Hence the sequence of states (of* Π) *which proves* $\pi \in \text{sols}(\Pi)$ *can be easily transformed to a sequence of states of* $(\Pi \triangleright \alpha)$ *which proves* $\pi_\alpha \in \text{sols}(\Pi \triangleright \alpha)$. *Thus* σ *is internally* α-*extensible.*

($\Leftarrow$) For every agent α_i, σ *is internally* α_i-*extensible and thus there is some local solution* π_i *such that* $\pi_i \in \text{sols}(\Pi \triangleright \alpha_i)$ *and* $\pi_i \triangleright \star = \sigma$. *When more than one local solutions exist, we can choose an arbitrary from them. Now we construct a solution* π *of* Π *from local solutions* π_i's *as follows. We split each* π_i *at the positions of public actions from* σ *and we join the corresponding internal parts of different plans together. Internal actions of different agents cannot interact through a shared fact (otherwise this fact would be public and these actions would be public too) and thus we can join different internal parts in any order, preserving only the order of actions of individual agents. Then we construct* π *of* Π *from* σ *by translating ids from* σ *to corresponding actions of* Π *and by adding the joined parts between corresponding public actions in* σ. *Clearly* $\pi \triangleright \star = \sigma$ *and* $\pi \in \text{sols}(\Pi)$. *Hence* σ *is internally extensible.* □

Example 9. We have seen previously that $\pi_2 \triangleright \star$ is internally *Truck*-extensible and also internally *Plane*-extensible. Hence we know that there is some solution of Π even without knowing π_0. Furthermore the proof of Theorem 1 shows how to reconstruct the solution. In contrast, we know that $\pi_1 \triangleright \star$ is not internally *Truck*-extensible and thus $\pi_1 \triangleright \star$ is not internally extensible.   ⊠

Some public plans of Π can be extended to a valid solution of Π but it might require inserting also public actions into σ. The following definition captures this notion which will be used in the following sections.

Definition 8. *Let public plan* σ *of* Π *be given. We say that* σ *is publicly extensible if there is public plan* σ' *of* Π *which is internally extensible and* σ *is a subsequence of* σ'.

Example 10. We have seen that $\pi_1 \triangleright \star$ is not internally extensible, however, it is still publicly extensible because it is a subsequence of $\pi_0 \triangleright \star$.

Similarly we can define that σ is *publicly* α-*extensible*. Projection solution $\pi \in \text{sols}(\Pi \triangleright \alpha)$ is called internally extensible (or publicly extensible) when the corresponding public plan $\pi \triangleright \star$ is so.

3   Confirmation Scheme

In this section we present a multiagent planning algorithm which effectively iterates over all solutions of one selected agent (*initiator*) in order to find such a solution which is internally extensible by all the other agents (*participants*). The confirmation algorithm provides a sound and complete multiagent planning algorithm (see Theorem 2).

Algorithm 1. Multiagent planning algorithm with iterative deepening.

input : multiagent planning problem Π
output : a solution π of Π when solution exists
Function MultiPlanIterative(Π) **is**

    $l_{max} \leftarrow 1$
    loop
        $\pi \leftarrow$ MultiPlan(Π, l_{max})
        if $\pi \neq \emptyset$ **then**
            | **return** π
        end
        $l_{max} \leftarrow l_{max} + 1$
    end

end

We assume that we have a separate agent capable of running planning algorithms for each agent mentioned in a given problem Π. Procedure MultiplanIterative from Algorithm 1 is the main entry point of our algorithms, both in this and the following sections. This procedure is initially executed by one of the agents called *initiator*. It takes a problem Π as the only argument and it iteratively calls procedure MultiPlan(Π, l_{max}) to find a solution of Π of length l_{max}, increasing l_{max} by one on a failure. In this way we ensure completeness of our algorithm because we enumerate the infinite set of all plans in a way that does not miss any solution. To simplify the presentation, we restrict our research only to those problems Π which actually have a solution, that is, $\mathsf{sols}(\Pi) \neq \emptyset$.

Algorithm 2 presents implementation of MultiPlan in the confirmation algorithm. We assume that SinglePlan($\Pi, \mathcal{F}, l_{max}$) implements a sound and complete classic planner which returns a solution of (an initiator projection of) Π of length l_{max} which is not in $\mathcal{F}$. Moreover we assume that SinglePlan always terminates and that it returns $\emptyset$ when there is no solution.

Initially, we set $\mathcal{F}$ to $\emptyset$. Then we invoke SinglePlan to obtain a solution of Π denoted as π. Afterwards, we ask the participant agents whether or not the public plan $\pi \triangleright \star$ is internally α-extensible. How participant agents fulfill this task is described in Sect. 5. When answers from all of the agents are affirmative then π is returned as a result. Otherwise π is added to the set of forbidden plans $\mathcal{F}$ and SinglePlan is called to compute a different solution.

The following states that the (public projection of the) plan returned by the confirmation algorithm is internally extensible to a solution of Π (*soundness*), and that the algorithm finds internally extensible solution when there is

---

Algorithm 2. MultiPlan($\Pi, l_{\max}$) in the confirmation scheme. Function SinglePlan($\Pi, \mathcal{F}, l_{\max}$) returns a plan of length $l_{\max}$ solving problem Π omitting forbidden plans from $\mathcal{F}$ or $\emptyset$ if there is no such plan. Method AskAllAgents($\pi \triangleright \star$) ask all agents α mentioned in the plan whether they consider the public plan $\pi \triangleright \star$ to be internally α-extensible and returns OK if all agents reply YES.

---

input : problem Π and a maximum plan length $l_{\max}$
output : a solution π of Π when solution exists
Function MultiPlan($\Pi, l_{\max}$) **is**

    $\mathcal{F} \leftarrow \emptyset$

    loop

        $\pi \leftarrow$ SinglePlan($\Pi, \mathcal{F}, l_{\max}$)

        if $\pi = \emptyset$ **then**

            | **return** $\emptyset$

        end

        $reply \leftarrow$ AskAllAgents($\pi \triangleright \star$)

        if $reply =$ OK **then**

            | **return** π

        end

        $\mathcal{F} \leftarrow \mathcal{F} \cup \{\pi\}$

    end

end

---

one (*completeness*). It is easy to construct a solution of Π given an internally extensible plan.

Theorem 2. *Let procedure* SinglePlan *in* MultiPlan *(Algorithm 2) be sound and complete. Then algorithm* MultiplanIterative *(Algorithm 1) with confirmation procedure* MultiPlan *is sound and complete.*

Proof. To prove soundness, let us assume that π is the result of the algorithm MultiPlanIterative. *Public plan $\pi \triangleright \star$ was confirmed by each agent α to be internally α-extensible. Thus, by Theorem 1, it is internally extensible and following the theorem proof we can reconstruct the whole solution of Π.*

Let us prove completeness. During each loop iteration in MultiPlan one plan is added to $\mathcal{F}$. There are only finitely many plans of length $l_{\max}$ and thus algorithm MultiPlan always terminates because SinglePlan is sound and complete. When Π is solvable, then some internally extensible solution π has to be eventually returned by SinglePlan at some point because SinglePlan is complete. This solution is then the result of MultiPlan (and hence the result of MultiPlanIterative) because, as a solution of Π, it has to be confirmed by all the participants.

4 Generating Plans Using Diverse Planning

In the previous section, we have assumed that function $\texttt{SinglePlan}(\Pi, \mathcal{F}, l_{\mathsf{max}})$ selects an arbitrary solution of Π of length l_{max} which is distinct from all the previous solutions stored in $\mathcal{F}$. In this section, we present an improved version of $\texttt{SinglePlan}$ which selects a solution based on evaluation of *qualities* of previously found solutions.

Section 4.1 defines the notion of plan metrics which are used to describe how much two plans differ. Based on these metrics we define in Sect. 4.2 a notion of the *relative quality* of a plan based on evaluation of previously considered solutions which were, however, rejected by at least one of the participant agents. Finally, Sect. 4.3 describes improved version of function $\texttt{SinglePlan}$.

4.1 Plan Metrics

While planning looks for a single solution of a problem, the goal of diverse planning is to find several *different* solutions. There are two main approaches to define how much two plans differ. Firstly, the difference of two plans can be defined by their membership to the same homotopy class [2]. Secondly, we can use a distance between plans. The distance can be defined either on *(i)* actions and their relations, or on *(ii)* states that the execution of a plan goes through, or on *(iii)* causal links between actions and goals [13]. In this paper, we use two metrics of the type *(i)*, that is, distance metrics defined on actions and their relations defined using their mutual positions in the plan.

Different Actions Metric. The *Different Actions Metric* counts the ratio of actions which are contained only in one of the plans. It is defined as follows. Let $\pi_0 \setminus \pi_1$ denote the plan π_0 with all the actions from π_1 removed.

$$\delta^{\mathcal{A}}(\pi_A, \pi_B) = \frac{|\pi_A \setminus \pi_B| + |\pi_B \setminus \pi_A|}{|\pi_A| + |\pi_B|}$$

This metric considers neither the ordering of actions nor the fact that some of the actions can be in a plan multiple times. Nevertheless, it is very simple for evaluation.

Levenshtein Distance Metric. The *Levenshtein Distance Metric* [8] is a general distance metric defined on two sequences. Let $\mathsf{trim}(\pi)$ be the plan π with the last action removed. Moreover let $\mathsf{diff}(\pi_A, \pi_B)$ be 1 if the last actions of π_A and π_B differ and 0 otherwise. Then the Levenshtein metric $\delta^{\mathcal{L}}(\pi_A, \pi_B)$ is defined as follows.

$$\delta^{\mathcal{L}}(\pi, \emptyset) = |\pi|$$
$$\delta^{\mathcal{L}}(\emptyset, \pi) = |\pi|$$
$$\delta^{\mathcal{L}}(\pi_A, \pi_B) = \min \begin{cases} \delta^{\mathcal{L}}(\mathsf{trim}(\pi_A), \pi_B) + 1 \\ \delta^{\mathcal{L}}(\pi_A, \mathsf{trim}(\pi_B)) + 1 \\ \delta^{\mathcal{L}}(\mathsf{trim}(\pi_A), \mathsf{trim}(\pi_B)) + \mathsf{diff}(\pi_A, \pi_B) \end{cases}$$

This metric describes how many changes using *elementary operations* have to be performed to convert one plan into another. The elementary operations are *add* an action into the plan, *remove* an action from the plan, and *replace* one action in the plan by another action.

4.2   Plan Quality Estimation

In Algorithm 2, the initiator agent generates its local solution π and asks participant agents to check whether $\pi \triangleright \star$ can be extended to a solution of their local problems. Each participant either accepts or rejects $\pi \triangleright \star$. Based on their replies, we can define the quality $\mathcal{Q}(\pi)$ of π as the ratio of the number of participants accepting $\pi \triangleright \star$ and the total number of participants.

$$\mathcal{Q}(\pi) = \frac{\#\,of\,participants\,accepting\,\pi \triangleright \star}{\#\,of\,all\,participants}$$

Hence the plan π with $\mathcal{Q}(\pi) = 1$ is accepted by all of the participants and the algorithm successfully terminates.

Once we have a plan π' whose quality has already been established, we can define a *relative quality* $\Delta(\pi, \pi')$ of an arbitrary π with respect to π' using a selected metric δ on plans as follows.

$$\Delta(\pi, \pi') = |\mathcal{Q}(\pi') - \delta(\pi, \pi')|$$

The relative quality $\Delta(\pi, \pi')$ is high when either $\mathcal{Q}(\pi')$ is high and π is close to π', or when $\mathcal{Q}(\pi')$ is low and π is distanced from π'. In other cases the value is close to zero.

Assume we have a set of plans $\mathcal{P}$ whose qualities have already been established. Then we can compute the relative quality $\Delta(\pi, \mathcal{P})$ of an arbitrary plan π with respect to $\mathcal{P}$ in several ways. In our work we work with the following two *quality estimators*.

Average Quality Estimator. The *average estimator* $\Delta^{\oslash}(\pi, \mathcal{P})$ is defined as the average of the relative qualities of π with respects to the plans from $\mathcal{P}$.

$$\Delta^{\oslash}(\pi, \mathcal{P}) = \frac{\sum_{\pi' \in \mathcal{P}} \Delta(\pi, \pi')}{|\mathcal{P}|}$$

Minimal Quality Estimator. The *minimal estimator* $\Delta^{\min}(\pi, \mathcal{P})$ is defined as the minimal relative quality.

$$\Delta^{\min}(\pi, \mathcal{P}) = \min_{\pi' \in \mathcal{P}} \Delta(\pi, \pi')$$

4.3 Generating Diverse Plans

During the execution of Algorithm 2, the initiator agent remembers the qualities $\mathcal{Q}$ of generated but rejected plans, that is, it remembers the qualities of all the plans from $\mathcal{F}$. We assume that $\mathcal{Q}$ is updated with every call to `AskAllAgents`. Additionally, the initiator computes the following statistics about actions.

$\mathcal{Q}(a)$ = average quality of plans containing a
$\mathcal{Q}(a, a')$ = average quality of plans containing a before a'

The function `SinglePlan` executed repeatedly by the initiator is described in Algorithm 3. It calls `DiversePlan` to generate a fixed number (n) of local solutions. Function `DiversePlan` works as follows. It starts by generating a solution candidate using roulette wheel selection [1] based on average action qualities $\mathcal{Q}(a)$. These actions are then presorted using statistics about action ordering $\mathcal{Q}(a, a')$. Note that two actions are swapped only if the difference of the statistics is larger then some threshold $\Delta^{\mathcal{Q}}$ (0.1 in our experiments). This ordering step allows algorithm to find the correct solution faster, but the price for that is lost of completeness of `SinglePlan` procedure.

Once a solution candidate is generated, the initiator α tests whether this sequence of actions is publicly α-extensible. If so, the solution is added to a set of diverse plans. This process is repeated until the required number of local solutions is found. In our implementation, this process is further extended and occasionally, instead of a roulette selection, those action which have not been used often are chosen. In this way the algorithm gathers further information about unused actions. Finally, function `SinglePlan` selects the diverse plan with the maximum relative quality.

5 Computing Plan Extensions

We have implemented the algorithms described in the previous sections taking advantage of several existing techniques and systems. An overall scheme of the architecture of our planner is sketched in Fig. 1. An input problem Π described in PDDL is translated into SAS using *Translator* script which is a part of Fast Downward[2] system. Our *Multi-SAS* script then splits SAS representation of the problem Π into agents' projections $\Pi \triangleright \alpha$ using user provided selection of public facts $P \triangleright \star$. Initiator then generates a solution to its own local problem. Participants are then requested to check whether they consider this extension to be α-extensible. When some of the participants finds the initiator solution is not α-extensible, then the initiator generates a new solution based on the responses of the participants.

The rest of this section demonstrates how the public and internal extension can be easily verified using any standard STRIPS planner.

In our algorithms, agents are asked whether a provided sequence of actions can be extended into a solution by adding other actions into the sequence.

---

[2] http://www.fast-downward.org/.

Algorithm 3. SinglePlan($\Pi, \mathcal{F}$) uses DiversePlan($\Pi, n, l_{\max}$) to generate n different solutions to the problem Π and then selects the best one using metric $\Delta(\pi, \mathcal{F})$. The generation of different plans is based on the roulette wheel selection by the quality evaluation received by other agents.

input   : classic STRIPS problem Π, the set $\mathcal{F}$ of forbidden plans, and a maximum plan length $l_{\max}$
output : a solution π of Π when solution exists
Function SinglePlan($\Pi, \mathcal{F}, l_{\max}$) **is**
    /* n is a constant   */
    $\mathcal{P}^{\mathrm{div}} \leftarrow$ DiversePlan($\Pi, \mathcal{F}, \mathsf{n}, l_{\max}$)
    $\pi \leftarrow \mathrm{argmax}_{\pi \in \mathcal{P}^{\mathrm{div}}}(\Delta(\pi, \mathcal{F}))$
    return π
end

input   : problem Π and n number of solutions
output : a set of diverse solutions
Function DiversePlan($\Pi, \mathcal{F}, n, l_{\max}$) **is**
    $\mathcal{P} \leftarrow \emptyset$
    while $|\mathcal{P}| < n$ **do**
        $A \leftarrow$ GetRandomActions(Π)
        $\pi' \leftarrow$ OrderActions(A)
        $\pi \leftarrow$ CreatePublicExtension(Π, π')
        if $\pi \neq \emptyset$ & $\pi \notin \mathcal{F}$ & $|\pi| \leq l_{\max}$ **then**
           $\mathcal{P} \leftarrow \mathcal{P} \cup \{\pi\}$
        end
    end
    return $\mathcal{P}$
end

Function GetRandomActions($\Pi, l_{\max}$) **is**
    $n \leftarrow$ RandomInt($1 \ldots \min(l_{\max}, |A|)$)
    $A \leftarrow \emptyset$
    while $|A| < n$ **do**
        $A \leftarrow A \cup \{a :$ roulette selection by $\mathcal{Q}(a)\}$
    end
    return A
end

Function OrderActions(A) **is**
    $\pi \leftarrow A$
    for $i = 2..|\pi|$ **do**
        if $\mathcal{Q}(\pi_i, \pi_{i-1}) - \mathcal{Q}(\pi_{i-1}, \pi_i) > \Delta^{\mathcal{Q}}$ **then**
           SwapActions(π_{i-1}, π_i)
        end
    end
    return π
end

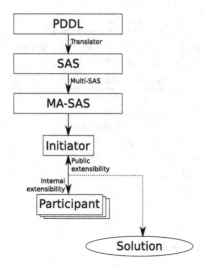

Fig. 1. Architecture of the planner.

Technically, this is similar to the planning problem with landmarks [3]. In this section we describe our algorithm to solve this problem. Based on this solution we describe how an initiator agent computes public extensions of a given sequence and how participant agents check whether a sequence of public actions is internally extensible.

Assume we are given a classic STRIPS planning problem $\Pi = \langle P, A, I, G \rangle$ together with a sequence $\sigma = \langle a_1, \ldots, a_n \rangle$ of actions build from the facts P. The planning problem with landmarks is the task to find a solution π of the problem $\langle P, A \cup \{a_1, \ldots, a_n\}, I, G \rangle$ such that σ is a subsequence of π, that is, that all the actions from σ are used in π in the proposed order. Note that an action a_i might or might be not in A.

Definition 9. *A planning problem with landmarks is a pair* $\langle \Pi, \sigma \rangle$ *where* $\Pi = \langle P, A, I, G \rangle$ *is a classic* STRIPS *problem and* $\sigma = \langle a_1, \ldots, a_n \rangle$ *is a sequence of actions build from the facts of* Π.

A solution π *of* $\langle \Pi, \sigma \rangle$ *is a solution of the classic* STRIPS *problem* $\langle P, A \cup \{a_1, \ldots, a_n\}, I, G \rangle$ *such that* σ *is a subsequence of* π.

We solve a planning problem with landmarks by translating $\langle \Pi, \sigma \rangle$ into a classic STRIPS problem Π^σ such that the solutions of Π^σ are in a direct correspondence to the solutions of the original problem with landmarks. We take a set P_{marks} of $n + 1$ facts distinct from P denoted as follows.

$$P_{marks} = \{mark_0, \ldots, mark_n\}$$

The meaning of fact $mark_i$ is that the landmark actions $a_1, \ldots, a_i$ has already been used in the correct order and that the action a_{i+1} can be used now.

We will ensure that only one fact from P_{marks} can hold in any reachable state. We will add $mark_0$ to an initial state and we will require $mark_n$ to be in the goal.

Definition 10. *Let* $\Pi = \langle P, A, I, G \rangle$ *and* $P_{marks} = \{mark_0, \ldots, mark_n\}$ *such that P and P_{marks} are distinct and* $\sigma = \langle a_1, \ldots, a_n \rangle$ *be given. For every action a_i let us define action b_i as follows.*

$$b_i = \langle\ \mathsf{pre}(a_i) \cup \{mark_{i-1}\},$$
$$\mathsf{add}(a_i) \cup \{mark_i\},$$
$$\mathsf{del}(a_i) \cup \{mark_{i-1}\}\ \rangle$$

The translation of the planning problem with landmarks $\langle \Pi, \sigma \rangle$ *into a classic* STRIPS *problem* Π^σ *is defined as follows.*

$$\Pi^\sigma = \langle\ P \cup P_{marks},\ A \cup \{b_1, \ldots, b_n\},\ I \cup \{mark_0\},\ G \cup \{mark_n\}\rangle$$

Basically we take action a_i and we add $mark_{i-1}$ to its preconditions and remove $mark_{i-1}$ when a_i is used. Moreover a use of a_i enables us to use the next action a_{i+1} from the list σ by adding $mark_i$ to the effects. It is easy to show the following property.

Lemma 1. *Let* $\langle \Pi, \sigma \rangle$ *be a planning problem with landmarks. When π is a solution of Π^σ then π with b_i's changed back to a_i's is a solution of $\langle \Pi, \sigma \rangle$. Moreover when there is a solution of $\langle \Pi, \sigma \rangle$ then there is a solution of Π^σ.*

Recall that every agent α is equipped with its local projection $\Pi \triangleright \alpha$ of problem Π, that is, a classic STRIPS problem defined as follows.

$$\Pi \triangleright \alpha = \langle P \triangleright \alpha, \mathsf{rel\text{-}actions}(\alpha), I \triangleright \alpha, G \rangle$$

The set $\mathsf{rel\text{-}actions}(\alpha)$ contains actions of agent α together with external actions. In Algorithm 3, the initiator agent, using $\mathtt{CreatePublicExtension}(\Pi, \pi')$, finds a solution of its local projection $\Pi \triangleright \alpha$ that has a given action sequence π' as a subsequence, that is, its public extension. The initiator can simply solve the planning problem with landmarks $\langle \Pi \triangleright \alpha, \pi' \rangle$ as shown in the following Theorem 3. Note that in this case the landmarks from π' are also in the set $A \triangleright \alpha$ of actions of $\Pi \triangleright \alpha$.

Theorem 3. *Plan π is publicly α-extensible to a solution of $\Pi \triangleright \alpha$ if and only if the planning problem with landmarks* $\langle \Pi \triangleright \alpha, \pi \rangle$ *is solvable. And moreover, the solution of planning problem with landmarks serves as a proof of the extensibility, and vice versa.*

Proof. It is quite straightforward to translate each plan π' proving public extensibility of plan π to a solution of the planning problem with landmarks $\langle \Pi \triangleright \alpha, \pi \rangle$, and vice versa.

In Algorithm 2, the participant agents are asked by the initiator from the call to $\mathtt{AskAllAgents}(\pi \triangleright \star)$ to establish whether $\pi \triangleright \star$ is internally α-extensible to a solution of $\Pi \triangleright \alpha$. The participant can simply check the solvability of the planning problem with landmarks $\langle a_1 \triangleright \alpha, \ldots, a_n \triangleright \alpha \rangle$ as shown in the following Theorem 4. Note that in this case the landmarks are not in $\mathsf{int\text{-}actions}(\alpha)$.

Theorem 4. *Plan* $\pi = \langle a_1, \ldots, a_n \rangle$ *is internally α-extensible to a solution of* $\Pi \rhd \alpha$ *if and only if the planning problem* $\langle P \rhd \alpha, \text{int-actions}(\alpha), I \rhd \alpha, G \rangle$ *with landmarks* $\langle a_1, \ldots, a_n \rangle$ *is solvable. Moreover, the solution of planning problem with landmarks serves as a proof of the extensibility, and vice versa.*

6   Experiments

For our experiments, we have designed the *Tool Problem* that allows us to observe a smooth transition in the complexity of the problem.

We focused our experiments on the following criteria: (1) comparison of different estimators and (2) an average number of iterations required to find a solution.

6.1   Tool Problem

In the *Tool Problem*, the goal is that each of N agents performs its public doGoal action as it is shown in Fig. 2. However, this action must be preceded by its internal useTool action first. Only the initiator agent can provide tools with the handTool action. Formally, there are N tools tool1, ..., toolN, and $N+1$ agents (the initiator and N participants). In the initial state, none of the participants has its tool and the initiator has all of them. However, the initiator does not know that the participants need them. One of possible solutions is as follows.

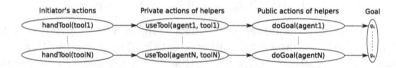

Fig. 2. A scheme of the Tool problem.

    1. handTool(initiator, tool1)
       ⋮
    N. handTool(initiator, toolN)
 N+1. useTool(participant1, tool1)
       ⋮
  2N. useTool(participantN, toolN)
2N+1. doGoal(participant1, tool1)
       ⋮
  3N. doGoal(participantN, toolN)

Other permutations of the plan also form a valid solution.

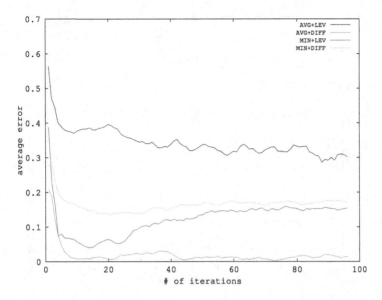

Fig. 3. Progress of an average error of plan qualities computed by different estimators for the *Tool Problem* with 10 tools.

6.2 Results

Let us present our results for the *Tool Problem* with 2, 4, 6, 8, 10, and 12 tools. Graphs in Figs. 3, 4, and 5 show the results of running our experiments 50 times.

Estimator Average Errors. We compare both estimators presented in this paper: Average Estimator (titled AVG in the graphs) and Minimal Estimator (MIN). Each estimator is tested with two different distance metrics: Different Action Metric (DIFF) and Levenshtein Distance Metric (LEV). Figure 3 demonstrates the progress of the estimators errors for the Tool Problem with 10 tools. Errors are computed from the average of 50 runs. As shown in the graph, Average Estimator with Different Action Metric converts quickly to very low error and thus it seems to be the best choice for this problem.

Figure 4 shows an average error for each estimator during first 80 iterations for different sizes of the Tool Problem. We can see that the Average Estimator with Different Action Metric again shows the lowest errors for all the cases, and furthermore, that its error decreases with increasing problem complexity.

Results for Tool Problem. Table 1 shows how many Tool Problems of different sizes has been solved during 50 runs using different plan generation techniques. We can see that most of the approaches perform better than a random generation of plans[3]. AVG+LEV again shows the best performance. Figure 5 shows more

---

[3] We have implemented a simple implementation of `SinglePlan` by translating a planning problem into a SAT problem instance and by calling an external SAT solver to solve it. It is easy to instruct a SAT solver to compute a solution different from previously found solutions.

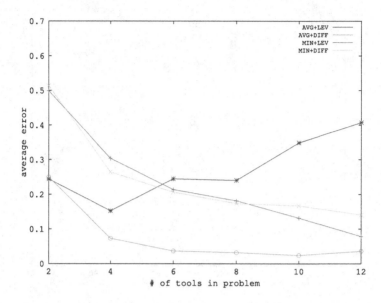

Fig. 4. Average errors of plan qualities computed from the first 80 iterations for the *Tool Problem* with a variable number of tools.

Table 1. Percentage of successfully solved instances of the Tools Problem for different number of tools. Comparison of a reference random plans generator (RND) and different combinations of estimators and plan distance metrics.

	2	3	4	5	6	7	8	9	10	11	12
RND	100 %	100 %	100 %	100 %	98 %	80 %	54 %	40 %	16 %	6 %	10 %
MIN+DIF	100 %	100 %	100 %	98 %	60 %	36 %	22 %	6 %	16 %	2 %	0 %
MIN+LEV	100 %	100 %	100 %	94 %	96 %	100 %	90 %	96 %	68 %	100 %	76 %
AVG+DIF	100 %	100 %	100 %	100 %	94 %	88 %	84 %	72 %	68 %	70 %	78 %
AVG+LEV	100 %	100 %	100 %	100 %	100 %	100 %	100 %	100 %	96 %	98 %	82 %

detailed distribution for its results in comparison to the baseline random generation. This graph shows a significant improvement over the baseline solution and that more complex cases of Tool Problem can be solved using this technique.

Results for IPC Problems. Classic planners are compared at the International Planning Competition with a well defined set of problems called *IPC problems*. However, most of these problems are by their nature single-agent problems and there is no standard way to convert them into a multiagent setting. Nevertheless, some of the problems are by their nature multiagent and fulfill all the requirements we have specified above in this paper. We have evaluated our algorithm on four domains: *Logistics, Openstacks, Rovers* and *Satellites. Logistics* problems contain two types of agents, *trucks* and *planes*, transporting packages between cities. *Openstacks* problems contain a *manager* agent who handles

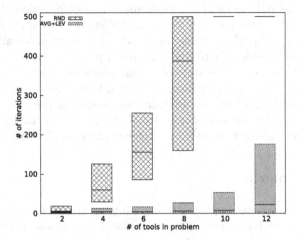

Fig. 5. The number of Generate-And-Test iterations needed to solve different sizes of the Tool Problem using random generation of plans (RND) and generation driven by the Average Estimator with the Levenshtein Distance Metric (AVG+LEV). Graph shows median (line in the rectangle) and 25 % and 75 % quantile (lower and upper bound of the rectangle) of the results.

product orders, and *manufacturer* agents who produce these products. The goal of *Rovers* problems is to plan actions for multiple robotic rovers on Mars that need to collect samples and transmit their data back to Earth via a shared base. *Satellites* problems contain from 1 to 12 *satellite* agents taking images in space.

Table 2. Number of iterations needed to successfully solve Openstacks, Rovers and Satellites problems from the IPC collection of planning problems. Problems marked by ∞ were not solved because the problem was too large for the test of public extensibility and FD did not finish in a reasonable time. The value 0 means that the problem contains only one agent and thus the solution was found immediately.

Problem no.	1	2	3	4	5	6	7	8	9	10	11	12	13	14	15	16	17	18	19	20	21–30
Openstacks 30/30	1	1	1	1	1	1	1	1	1	1	1	1	1	1	1	1	1	1	1	1	1
Rovers 7/20	0	0	10	15	1	1	∞	∞	∞	∞	∞	∞	∞	5	∞	∞	∞	∞	∞	∞	N/A
Satellites 9/20	0	0	1	∞	∞	1	∞	1	∞	1	∞	1	∞	∞	∞	∞	∞	1	3	∞	N/A

Table 2 shows the results for domains which were solvable by presented algorithm. None of the *Logistics* problem was solved within given 5 min time-limit. In contrast, all the problems of *Openstacks* were solved in first iteration. This is caused by the structure of the domain where every internally α-extensible solution is internally extensible. This is not always true for *Rovers* and *Satellites* domains and thus only some of the problems were solved during the first iteration, while the others require more iterations. When the first generated plan was not a solution of the problem then the search for a solution usually timeouted because it requires a planner to find out that a problem has no solution. This constitutes a challenge for the state-of-the-art planners which usually performs best on problems which actually have a solution. When there is no solution then the planners usually get stuck in an exhaustive search of the whole plan space. Nevertheless, our planner was able to solve quickly few of harder instances of the problem, even faster than other multiagent planners [9, Table 1].

7   Final Remarks

We have proposed a novel approach to planning for MA-STRIPS problems based on the Generate-And-Test principle and initiator–participant protocol scheme. We have experimentally compared various combinations of plan quality estimators and plan distance metrics improving efficiency of the plan generating approach. Additionally, we have validated a principle of planning with landmarks by compilation to classic planning problem used as the testing part of the planner. The results show that the principle is viable and the best combination of estimator and metric for the designed domain is averaging with action difference metric.

In future work, we plan to test the planner in more planning domains, as it is from the beginning designed as domain-independent and reinforce the plan generation process by elements of backtracking search. Additionally, the approach hinges on efficient solving of plan-(non)existence problems with landmarks (the plan extensibility problem), therefore we will analyze how to improve on that as well.

Acknowledgements. This research was supported by the Czech Science Foundation (grants no. 13-22125S and 15-20433Y).

References

1. Bäck, T.: Evolutionary Algorithms in Theory and Practice: Evolution Strategies, Evolutionary Programming, Genetic Algorithms. Oxford University Press, Oxford (1996)
2. Bhattacharya, S., Kumar, V., Likhachev, M.: Search-based path planning with homotopy class constraints. In: Felner, A., Sturtevant, N.R. (eds.) SOCS. AAAI Press (2010)
3. Brafman, R., Domshlak, C.: From one to many: planning for loosely coupled multi-agent systems. In: Proceedings of ICAPS 2008, vol. 8, pp. 28–35 (2008)
4. Decker, K., Lesser, V.: Generalizing the partial global planning algorithm. Int. J. Intell. Coop. Inf. Syst. **1**(2), 319–346 (1992)
5. Doherty, P., Kvarnström, J.: TALplanner: a temporal logic-based planner. AI Mag. **22**(3), 95–102 (2001)

6. Durfee, E.H.: Distributed problem solving and planning. In: Weiß, G. (ed.) A Modern Approach to Distributed Artificial Intelligence, Chap. 3. The MIT Press, San Francisco (1999)

7. Fikes, R., Nilsson, N.: STRIPS: A new approach to the application of theorem proving to problem solving. In: Proceedings of the 2nd International Joint Conference on Artificial Intelligence, pp. 608–620 (1971)

8. Levenshtein, V.: Binary codes capable of correcting deletions, insertions and reversals. Soviet Physics Doklady **10**, 707–710 (1966)

9. Nissim, R., Brafman, R.I.: Multi-agent A* for parallel and distributed systems. In: Proceedings of AAMAS 2012, pp. 1265–1266. Richland, SC (2012)

10. Nissim, R., Brafman, R.I., Domshlak, C.: A general, fully distributed multi-agent planning algorithm. In: Proceedings of AAMAS 2010, pp. 1323–1330. Richland, SC (2010)

11. Pellier, D.: Distributed planning through graph merging. In: Filipe, J., Fred, A.L.N., Sharp, B. (eds.) ICAART, vol. 2, pp. 128–134. INSTICC Press (2010)

12. Richter, S., Westphal, M.: The LAMA planner: guiding cost-based anytime planning with landmarks. J. Artif. Int. Res. **39**(1), 127–177 (2010)

13. Srivastava, B., Nguyen, T.A., Gerevini, A., Kambhampati, S., Do, M.B., Serina, I.: Domain independent approaches for finding diverse plans. In: Veloso, M.M. (ed.) IJCAI, pp. 2016–2022 (2007)

14. Torreño, A., Onaindia, E., Sapena, O.: An approach to multi-agent planning with incomplete information. In: ECAI, pp. 762–767 (2012)

15. Tožička, J., Jakubův, J., Durkota, K., Komenda, A., Pěchouček, M.: Multiagent planning supported by plan diversity metrics and landmark actions. In: Proceedings ICAART 2014. SciTePress (2014)

Concurrent and Distributed Shortest-Path Searches in Multiagent-Based Transport Systems

Max Gath$^{(\boxtimes)}$, Otthein Herzog, and Maximilian Vaske

Technologie-Zentrum Informatik und Informationstechnik Institute
for Artificial Intelligence, Am Fallturm 1, 28359 Bremen, Germany
{mgath,herzog,zottel}@tzi.de
http://ai.uni-bremen.de

Abstract. The Fourth Industrial Revolution and the consequent integration of the Internet of Things and Services into industrial processes increase the requirements of transport processes. Customer demanding same-day deliveries, shorter transit-times, individual qualities of shipments, and higher amounts of small size orders raise the complexity and dynamics in logistics. In these highly dynamic environments, multiagent systems (MAS) and multiagent-based simulation (MASB) offer a suitable approach to handle the complexity and to provide the required flexibility, robustness, as well as customized behavior. This article focuses on the impact and the relevance of shortest-path queries in MAS and MABS. It compares the application of state-of-the-art algorithms and investigates different modeling approaches for efficient and concurrent shortest-path searches. The results prove that the application of a highly efficient algorithm such as hub labeling with contraction hierarchies is an essential key component in the agent-based control of dynamic transport processes. Moreover, the results reveal that choosing a modeling approach which slightly restricts the agents' autonomy increases significantly the runtime performance without losing the advantages of multiagent systems. This allows for applying MAS to solve large scale real-world transport problems and for performing MABS with low hardware requirements.

Keywords: Multiagent systems · Multiagent-based simulation · Shortest-path searches · Parallel and distributed algorithms · Planning and scheduling · Hub labeling with contraction hierarchies

1 Introduction

The so-called goods structure effect refers to a change of the economic and logistic structure: The production of bulk goods which are transported in large quantities by bulk cargo transport has been decreased, while the amount of individualized high-end products has been increased. This trend is aggravated by the so-called Industry 4.0 – the Fourth Industrial Revolution and the consequent integration of the Internet of Things and Services in production and logistics processes based on Cyber Physical Production Systems.

© Springer-Verlag Berlin Heidelberg 2015
N.T. Nguyen et al. (Eds.): TCCI XX, LNCS 9420, pp. 140–157, 2015.
DOI: 10.1007/978-3-319-27543-7_7

As a result, there is a much higher amount of small size shipments, which have to be delivered within guaranteed time windows and probably within a few hours. The demanding customer requirements and the growing cost pressure in the logistic sector thus forces logistic transport service providers to optimize the efficiency of their processes. Multiagent systems (MAS) can be used to solve complex, dynamic, and distributed problems [36] in which agents are a natural metaphor for physical objects and actors [28, p. 7]. Consequently, multiagent systems are an adequate technology for the modeling and the optimization of logistic processes. It has been shown that their application to logistics increases the efficiency as well as the service quality and contributes to reduce the costs significantly. Table 1 provides an overview of multiagent systems which were developed for transport logistics.

Multiagent-based simulation (MABS) combines concepts of multiagent systems and simulation. Applying MABS allows for the analysis of MAS before their deployment to real world processes. Thus, it is possible to investigate the impact of potential changes, to calculate expected benefits, and to identify risks in advance that may arise by switching to new processes and by the integration of new technologies such as MAS. This is especially relevant in scenarios where the quality of the results depends on the outcome and/or sequence of agent negotiations that cannot be predicted in advance [27]. Moreover, MABS allows for precise scenario investigations and strategic analyses. For instance, affects of new pricing models or the impact of economic cycles and natural disasters on the supply chain can be determined.

In all multiagent-based approaches for transport logistics, it is essential to compute the distances between cities (stops). For instance, the agents which represent vehicles must often solve a Traveling Salesman Problem (TSP) [15] to determine their routes and to calculate the cost for their proposals in agent-based negotiations. Therefore, a distance-matrix between all stops is required as input for the TSP solver. In the established and often applied benchmarks for transport problems, the required distances between locations are Euclidian distances. These can be determined with low computational requirements. However, shortest-path searches on real-world infrastructure networks are cost-intensive operations. Therefore, efficient shortest-path searches are essential for transport processes and especially for multiagent-based approaches. Despite the considerable importance of shortest-path searches, none of the multiagent-based approaches presented in Table 1 focuses on this problem.

The goal of this research is to optimize the runtime performance of MAS and MABS in transport logistics. Section 2 presents the MABS framework PlaSMA, which has been developed for simulations of logistic processes. Next, Sect. 3 presents our developed multiagent system for the optimization of planning and scheduling processes. The system is used as reference system in the following investigations. Section 4 compares the application of well-established and high speed shortest-path algorithms. It presents the implemented algorithms and investigates the effects of shortest-path computations in multiagent-based negotiations. Section 5 focuses on the parallel application of a state-of-the-art *hub*

Table 1. The table depicts authors who provide multiagent systems for transport logistics.

Authors	Reference
Fischer, Müller, and Pischel	[14]
Bürckert, Fischer, and Vierke	[8]
Thangiah, Shmygelska, and Mennell	[39]
Perugini, Lambert, Sterling, and Pearce	[37]
Dorer and Calisti	[11]
van Lon, Holvoet, Vanden Berghe, Wenseleers, and Branke	[32]
Kohout and Erol	[30]
Leong and Liu	[31]
Mes, van der Heijden, and van Harten	[35]
Barbucha and Jedrzejowicz	[4]
Zhenggang, Linning, and Li	[42]
Himoff, Skobelev, and Wooldridge	[25]
Himoff, Rzevski, and Skobelev	[26]
Glaschenko, Ivaschenko, Rzevski, and Skobelev	[22]
Mahr, Srour, de Weerdt, and Zuidwijk	[34]
Vokřínek, Komenda, and Pěchouček	[40]
Kalina and Vokřínek,	[29]
Gath, Herzog, and Edelkamp	[17]

labeling algorithm with *contraction hierarchies*. Several modeling approaches are compared and discussed. The goal is to increase the performance of the decision making process of agents wrt. the runtime and memory consumption by an efficient implementation of shortest-path searches and the use of an adequate agent model. This is crucial for the implementation of autonomous control in real-world transport processes as well as for large scale logistic multiagent simulations. Finally, Sect. 6 concludes the article and gives future research perspectives.

2 The PlaSMA Simulation Platform

The PlaSMA simulation platform [41] is an agent-based event driven simulation platform that has been designed for modeling, simulation, evaluation, and optimization of planning and controlling processes in logistics. It extends the FIPA-compliant Java Agent DEvelopment Framework (JADE) [7] for agent communication and coordination. PlaSMA provides discrete time simulations, which allow for precise simulations of processes with small simulated time intervals (with intervals of at least 1ms). Furthermore, it ensures correct synchronization and reproducibility [19]. For instance, the simulation framework guarantees, that message transfer consumes simulated time, because transferring messages consumes physical time in real-world processes as well. Consequently, the consistency of each agent (e.g., to ensure that no agent receives messages from the future and all the agents' knowledge is consistent at a certain point of simulated time) is also guaranteed by a conservative synchronization mechanisms (cf. [19] for more details). The time model adequacy is assured by a parameter which controls the maximum and minimum simulated time interval for the synchronization. Thus, PlaSMA is capable to simulate scenarios that require fine-grained and coarse time discretization as well.

Moreover, it supports the integration of real-world infrastructures by the import of geographic information from OpenStreetMap and of timetable information (e.g., for bus lines or tram lines), which matches the standards of the Association of German Transport Companies (VDV) [23]. The transport infrastructure is represented by a directed graph where edges represent ways such as waterways, rails, and roads while nodes represent traffic junctions that connect edges with each other. The type of the road (e.g., highway, inner city road, or pedestrian way) including its properties (e.g., speed limits, exact distances, and oneway restrictions) is further specialized automatically by processing the respective information provided by the OpenStreetMap (see: http://openstreetmap.org) dataset. Thus, PlaSMA allows for modeling fine-grained infrastructures with road sections whose speed limits are differing. Shortest-path searches on large real-world infrastructures are some of the most cost-intensive operations in logistic scenarios. A pre-computation of a whole distance matrix quickly exceeds the available memory. Especially in dynamic environments a regular re-computation of a whole distance matrix is too time-consuming and impossible on large infrastructures. Therefore, this paper focuses on an efficient implementation and modeling approaches of shortest-path searches in multiagent systems.

In order to reliably simulate industrial and transport processes, PlaSMA is capable of incorporating process data of cooperating companies and partners, e.g., customer orders or service requests, directly into the simulation platform. This allows for a precise analysis of real logistic processes with low costs. Batch-runs, process visualization, as well as automated measurements of individually defined performance indicators allow for fast and significant process evaluations. Figure 1 shows the graphical user interface of PlaSMA. The software can be downloaded from http://plasma.informatik.uni-bremen.de.

3   A Multiagent System for the Optimization of Transport Processes

This section presents our developed multiagent system for transport logistics, which is used as the reference system for the investigations in Sects. 4 and 5. In our approach agents represent transport vehicles and orders. The agents differ in their individual properties, e.g., represented vehicles vary in their capacities, work schedules, and speed limits. Similarly, each *order agent* carries the unique

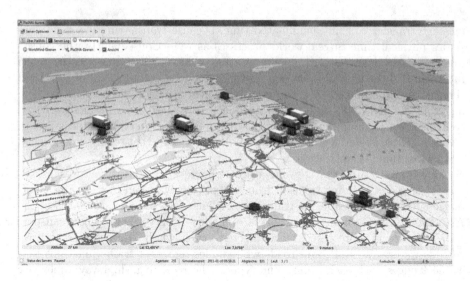

Fig. 1. The graphical user interface of the PlaSMA simulation platform.

characteristics of its represented shipment such as the pickup and delivery location, weight, value, time windows, and premium service constraints. The goal of order agents is to find a proper transport service provider for transporting the shipment from the depot to the destination (or vice versa) within given time windows. *Vehicle agents* negotiate and communicate with order agents to maximize the number of carried shipments while satisfying all relevant constraints and premium service priorities.

The system starts with a rough planning step by applying a k-means algorithm [33]. In contrast to a static mapping from postal codes to tours, the algorithm assigns only effectively arrived orders to available vehicles by grouping orders in nearby districts. Consequently, the system is able to react to daily as well as seasonal fluctuations. After the rough planning step, each vehicle agent starts a detailed planning process. On the one hand, the vehicle agent considers the represented truck's capacity, the driving times, which depend on the type of the road and the respective speed limits, as well as the individual capacities of the shipments such as the weight, priority, time windows, handling times, and the pickup or delivery location. On the other hand, the agent optimizes the objective functions to reduce cost and determine efficient solutions. For instance, the vehicle identifies the shortest path for visiting all stops. As a result, the agent solves a selective TSP which is NP hard [9]. For this purpose, the vehicle agent applies an optimal solver for small size problems [12]. In order to solve larger problems, a Nested Monte Carlo approach with Policy Adaptation (also called Nested Rollout Policy Adaptation) was developed [13]. Both solvers require the distance matrix, which must include all the cities (stops) that have to be visited and the current position of the vehicle. Thus, several shortest-path searches are applied for the calculation of this distance matrix. Especially in dynamic environments with changing traffic conditions and in scenarios in which orders must be scheduled during operations, it is infeasible to precalculate all the distances offline. This may only be an adequate solution for static problems if all service requests are known in advance. After the detailed planning step of each vehicle agent, several orders may not be serviced by a vehicle. Thus, the responsible agent acts in the same way like agents representing dynamically incoming orders: The agent sends a transport request to available vehicle agents and starts a new negotiation. The vehicle agents compute proposals by determining their additional cost for transporting the shipment. In order to schedule new orders also while transporting other shipments, the agent considers all relevant changes of the environment and its internal state, e.g., already loaded shipments and the current position of the vehicle. The computed cost is sent back to the order agent that chooses the transport provider with the least cost. If it is not possible to satisfy the orders' requirements, a refuse message is sent by the vehicle agent. To transport a premium service instead of conventional orders, or another premium service with less cost, already accepted orders (that have not been boarded yet) may not be included in the new plan and must be rescheduled. Affected order agents negotiate with other transport service providers again. The agent models consider concurrency aspects within negotiations, the dynamics of the environment, as well as the interdependencies between planning and the execution of an existing plan. Details of the proactive and reactive agent design are provided by Gath et al. (2013) [18] and by Gath et al.(2014) [17].

4   The Impact of Shortest-Path Searches

Three well-established shortest-path algorithms are implemented for the first investigation: the classical Dijkstra algorithm [10], an implementation of the A* algorithm, which both use radix-heaps [24] for an efficient graph representation, as well as a shortest-path algorithm which combines hub labeling with contraction hierarchies. The section starts with an overview of the implemented algorithms. Next, it presents the experimental setup, shows the results, and discusses the impact of the different algorithms in the multiagent-based transport system described in Sect. 3.

4.1   The Dijkstra Algorithm

Let N denote a set of nodes, E a set of edges, and $dist : E \rightarrow \mathbb{R}$ the distance function of an edge. The Dijkstra shortest-path algorithm [10] is probably the best known and most frequently applied algorithm for the computation of a shortest path $P = (e_0, ..., e_l)$ with $e \in E$ and with the minimum distance $\min_P \sum_{e \in P} dist(e)$ between two nodes $s, t \in N$. Nodes are labeled as visited, reachable, or unvisited and contain a reference to the predecessor node and the distance to node s. At the beginning, all nodes are unvisited and have an infinite distance to s. While reachable nodes exist and the target t is not visited, the algorithm performs the following steps: Firstly, it chooses the reachable nearest node c. Then, node c is labeled as visited and all unvisited nodes n which are connected to an outgoing edge are evaluated next. Let d_n denote the distance from s to n. If n is reachable and d_n is smaller than its currently saved distance, node n is marked as reachable with distance d_n and with the predecessor node c. After termination, the distance of t denotes the shortest distance from s to t (if there exists a path from s to t).

4.2   The A* Algorithm

The A* algorithm is similar to the Dijkstra algorithm, but applies an additional heuristic which underestimates the real cost from the processed node to the target to push the search in the right direction and to avoid the expansion of nodes which are not part of the shortest path. For instance, on road networks the Euclidian air distances may be used as a valid heuristic. We implemented a memory-efficient version of the A* algorithm which is based on radix heaps. As a result, each node can be processed in constant time and space. Details of this algorithm are provided by Greulich et al. (2013) [24].

4.3   Hub Labeling with Contraction Hierarchies

Since 2011, hub labeling algorithms [1] in combination with contraction hierarchies (CH) [20] are known to be the most efficient shortest-path algorithms. For instance, shortest-path queries on the whole transport network of Western Europe are processed in less than a millisecond [2, p. 34].

The idea of distance labeling algorithms is that the distance between two nodes is only determined by the comparison of their assigned labels, which are ideally computed offline (cf. Fig. 2). Therefore, search queries on the pre-computed labels can efficiently be performed online. In our implementation, the hub labels contain a list with references to multiple other nodes (the hubs). Within the construction process of the labels, the so-called *cover property* has to be satisfied, which means that both labels of any two vertices s and t must contain the same vertex that is on the shortest $s - t$ path [2, pp. 25–26]. The cover property guarantees that all shortest paths in a graph can be determined by the labels of the source and target nodes. The challenge is to create memory-efficient labels that satisfy the cover property.

Applying the labeling algorithm on nodes which are saved in a CH, allows for memory-efficient label representations. In order to build the CH, the original graph g is extended to a larger graph g' which contains direct shortcuts between nodes instead of shortest paths in g. The algorithm iterates over all nodes and saves each node in the next higher level of the hierarchy. In this process, it calculates possible shortest-path shortcuts to other nodes. Therefore, the current node is considered to be removed from the graph and it is checked if all other shortest paths would still be included within the graph without this node. If a shortest path originally passed the *removed* node, a new shortcut is created to retain this shortest path. The general steps are shown in Fig. 3.

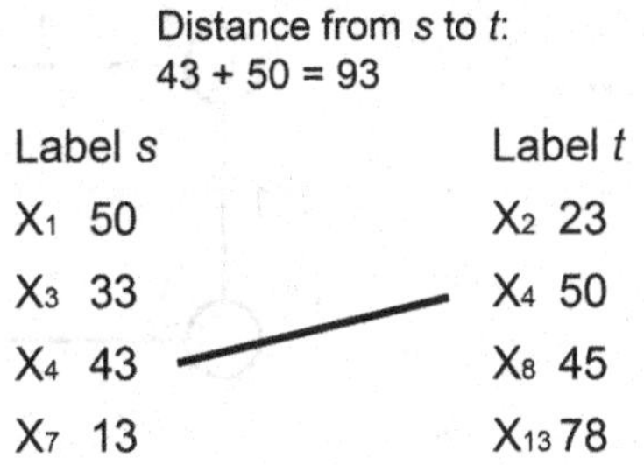

Distance from s to t:
43 + 50 = 93

Label s		Label t	
X_1	50	X_2	23
X_3	33	X_4	50
X_4	43	X_8	45
X_7	13	X_{13}	78

Fig. 2. This example shows how the distance between a start node s and a target node t is determined by the comparison of their labels.

The performance of the algorithm depends highly on the sequence of nodes in which they are added to the CH [20]. An optimal sequence minimizes the search space for the optimal number of shortcuts. As this problem is NP-hard [6], Geisberger et al. (2008) [20, pp. 322–324] suggest to use the following approach. To determine the next node that will be processed, all unprocessed nodes are sorted by a priority value. The node with the highest priority is processed next. The priority value of a node is mainly computed by the edge difference between the current graph and the graph containing the shortcuts that result from processing that node. Some priorities have to be updated continuously after adding a new node to the CH, because in every iteration the graph might be extended by a new shortest path. Due to the fact that the computation of the priorities is a cost-intensive operation, the value is estimated. The better the sequence of the iterated/selected nodes is, the less shortcuts are determined, and the more efficient is the memory consumption and the search on the CH. In addition, there are also approaches which can be applied to time-dependent graphs [5] or to dynamically changing graphs [21].

Next, the hub labels are computed on the basis of the CH and the extended graph g'. This process starts at the highest level of the CH. For each level (node) a label is created. The label contains all references and information about the shortest distance to nodes saved in higher levels. Figure 4 gives an example for the creation of hub labels on the basis of a CH and a graph g'. Further optimization techniques to reduce the memory are not implemented yet, but provided by Abraham et al. (2012) [2].

4.4 Experimental Setup

The goal is to investigate the impact of shortest-path searches on the runtime in multiagent-based logistic transport processes making use of the multiagent system presented in Sect. 3. Therefore, several simulations of exactly the same scenario were performed with the three different shortest-path algorithms within PlaSMA (cf. Sect. 2). The simulated scenario is similar to the one, which is presented by Gath et al. (2013) [18]. The underlying real-world transport

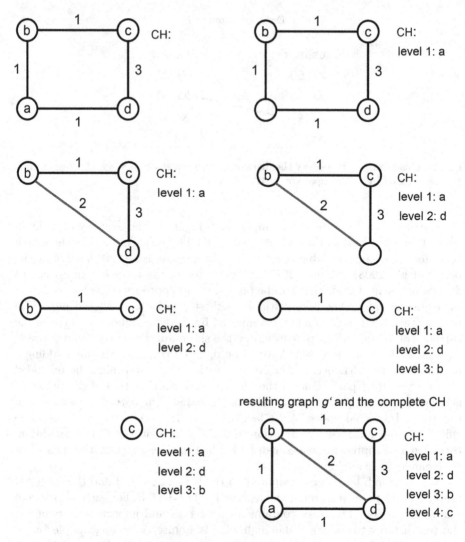

Fig. 3. In order to build the CH and the graph g', the algorithm iterates over all nodes and saves each node in the next higher level of the hierarchy. Therefore, it removes the considered node from the graph. If a shortest path originally passed the *removed* node, a new shortcut is created in g' to retain this shortest path.

infrastructure, which is imported from OpenStreetMap[1], contains 85, 633 nodes and 196, 647 edges. All experiments were performed on a laptop computer with an Intel quad-core i72620-M CPU/2.7 GHz and 16 GB RAM.

---

[1] http://www.openstreetmap.org (cited: 22.04.15).

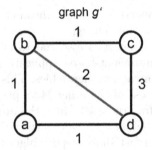

graph g'

CH: hub labels:

level 1: a

level 2: d

level 3: b

level 4: c   c: c,0

CH: hub labels:

level 1: a

level 2: d

level 3: b   b: b,0; c;1

level 4: c   c: c,0

CH: hub labels:

level 1: a

level 2: d   d: d,0; b,2; c,3

level 3: b   b: b,0; c;1

level 4: c   c: c,0

CH: hub labels:

level 1: a   a: a,0; b,1; d,1; c,2

level 2: d   d: d,0; b,2; c,3

level 3: b   b: b,0; c;1

level 4: c   c: c,0

Fig. 4. The figure shows an example how the hub labels are created on the basis of a CH and a graph g'. Starting at the highest level, for each level (node) a label is created. The label contains all references and information about the shortest distance to nodes saved in higher levels.

4.5 Results

Figure 5 compares the physical time which is required for the simulation of a scenario in PlaSMA. It clearly shows that the shortest-path algorithm significantly affects the runtime of the multiagent-based dispatching approach even on relatively small infrastructures investigated in this experiment[2].

4.6 Discussion

Although the Dijkstra and A* algorithm are well-established and algorithms applied often, Fig. 5 shows that the shortest-path problem, whose complexity is in P, has a more significant impact to the system runtime than solving the NP-hard TSP with a high-performance algorithm. Thus, if the well-established

---

[2] In the experiment a single routing agent maintains a the shortest-path algorithm (see next Section).

Dijkstra or A* algorithm is applied, it is clearly the most time-intensive operation of the multiagent-based dispatching system.

Consequently, this proves that it is essential to apply a high-performance shortest-path algorithm in multiagent-based applications for transport logistics such as hub labeling with contraction hierarchies. This is not only true for the implemented MAS, but for most of the other MAS presented in Table 1, which are applied on real-world infrastructures, since the order agents have to compute a distance matrix first before they can solve the VRP-like sub-problems. Only multiagent systems with fast shortest-path algorithms have an acceptable runtime performance in real-world scenarios. In addition, consuming less time to compute the distance matrix allows for increasing the number of negotiations running concurrently. This enables the agents to validate more options and consequently optimizes the overall solution quality.

5   Comparison of Different Agent Modeling Approaches

In general, an agent has only two options to acquire shortest-path information. On the one hand, the agent can compute the shortest paths by itself. On the other hand, the agent might ask a service provider agent (a so-called routing agent) which receives a routing request (from a so-called consumer agent), computes the shortest path, and finally sends the result back to the respective consumer agent. The goal of the second investigation is to determine an adequate way of modeling for a scenario in which numerous routing requests have to be answered immediately (e.g., to compute distance matrices of cities). Especially if the agents apply the hub labeling algorithm in combination with CH, it is not sufficient to consider the performance to handle search queries online, but also necessary to include the time for the creation of the hub labels and of the CH as well as the amount of memory, which is required to save all the labels.

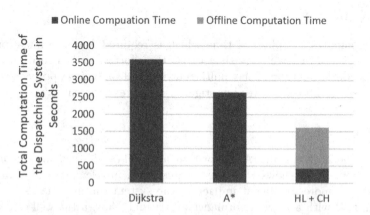

Fig. 5. A comparison of the runtime of multiagent-based transport simulations. The only difference between the scenarios is the applied shortest-path algorithm.

In the first case, each agent has its full autonomy and maintains its own algorithm. However, each agent has to build and save the hub labels as well. This is time- and memory-intensive. In the second case, the agents outsource the memory and time consuming shortest-path operations to one or several routing agents. As a result, only the routing-agent must build and save the hub labels. In this case, the optimal number of parallel running routing agents must be determined.

Beside the above mentioned options, the Java programing language allows another approach that technically slightly restricts an agent's autonomy. It is possible to build the hub labels by a single agent and save them in a static variable. While classical routing algorithms such as the Dijkstra algorithm manipulate the graph by saving distance information at the nodes to compute the shortest paths, the hub labeling algorithm performs read-only operations on the labels instead. Thus, the agents can directly access this static variable and perform the routing requests by their own in parallel. Depending on the computer architecture the multiagent system is running on, these operations are performed also physically concurrently. However, this slightly violates the agent's autonomy, because all agents (running on the same Java Virtual Machine (JVM)) share the same component. Thus, they are technically not fully independent of each other.

5.1 Experimental Setup

The following investigation focuses on scenarios simulated in PlaSMA. In these scenarios 1,000,000 routing requests of several agents must be answered immediately (e.g., to compute distance matrices of cities/stops). To satisfy real-world requirements, the underlying transport infrastructure is the road network of the country of Lichtenstein with 3,607 nodes and 8,401 edges. For the evaluation in reasonable time on conventional hardware, this area with a restricted number of nodes and edges is choosen, because it allows to pinpoint significant results by measuring average values of 10 runs in each setting. Nevertheless, the algorithm has successfully been applied to larger infrastructures with more than 300,000 edges and 200,000 nodes. The 1,000,000 search queries are requested by 50 agents. Thus, each agent asks for 20,000 shortest paths. The simulation is started with an agent, which generates 20,000 search queries with randomized start and end nodes for each of the 50 consumer agents which are created. In order to guarantee the reproducibility of runs and an accurate comparison, the random seed of the random number generator is fixed in each experiment.

In Scenario 1, each consumer agent applies its own shortest-path algorithm. The agents start a pre-processing step to build up the hub labels as well as the CH and process all their queries by themselves. In Scenario 2, the shortest-path algorithm is implemented by the *Singleton Pattern* [16, pp. 127]. In this scenario, there exists only a single instance of the algorithm on each JVM. All consumer agents process their queries by themselves, but operate on the same

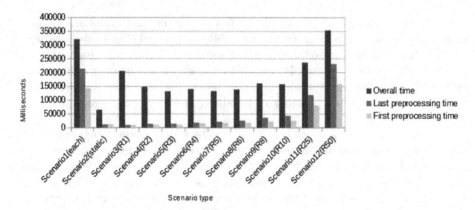

Fig. 6. The average simulation time and the average time required to create the hub labels (of 10 runs) of the simulated scenarios.

instance of the implemented algorithm saved in a static variable[3]. In Scenario 3–12, between 1 and 50 routing agents are created. They maintain their own shortest-path algorithm and receive all the queries from the consumer agents by a FIPA-compliant ACL-Message. Next, the routing agents compute the shortest path of the assigned requests and send an ACL-Message with the answer back to the agent. The assignment of search requests to routing agents is uniformly distributed. The reproducibility of this assignment is also ensured by applying a fixed random seed. All the simulations run on a notebook computer with an Intel quad-core i5-2500k processor, Windows 7 64 bit, and 16 GB RAM.

5.2 Results

In each scenario, four performance indicators were measured. The most significant is the total (physical) simulation time. Moreover, two performance indicators determine the time required for pre-processing. This is the earliest time an agent ends its pre-processing step and the elapsed time for all the agents to finish building the hub labels and the CH. The results are shown in Fig. 6. The measured performance indicators are average values of 10 runs. As the shortest-path algorithm only performs read-only operations, the memory requirements of the whole MAS increase proportionally to the number of graph instances, which are created and saved concurrently by all agents. This is obvious, but also rechecked in this investigation.

5.3 Discussion

The results show that providing the *full* technical autonomy of agents requires a higher memory utilization and longer runtime. Each agent must create and

---

[3] Note that PlaSMA extends JADE. Thus, each static variable is only visible to the JVM. In distributed simulations on multiple machines, each machine requires its own static routing algorithm.

maintain its own routing algorithm. On a quad-core architecture, this process cannot be parallelized physically to 50 agents, but must be performed sequentially. Thus, sourcing out the cost-intensive operations to routing agents is reasonable. As a result, the pre-processing time as well as the total simulation time is reduced. In JADE, each agent has its own process. However, if there is just a single routing agent, the hardware utilization is only about 25% on a quad-core processor, because all the search queries are executed by a single agent on a single core. Consequently, the pre-processing time and the total simulation time can be reduced by physically concurrently running routing agents as long as the number of routing agents is lower or equal to the number of available cores. To this end, the search queries are answered in parallel and the hardware utilization is increased. If the number of routing agents exceeds the amount of cores, the pre-processing for the creation of the hub labels cannot be performed in parallel anymore and the total time for pre-processing and simulation is increased. If each consumer agent has its own routing agent, the performance is even higher compared to the case that each agent has its own algorithm. This is explained by the fact, that the message transfer within the multiagent system consumes additional time.

Moreover, the results show that the shortest running time (and also the lowest memory utilization) is reached by the static implementation of the routing algorithm. This is only possible if the shortest-path algorithm performs read-only operations during the search such as hub labeling with a CH. It is not achievable with the classical Dijkstra or A* algorithm. Even if we compare the result of this modeling approach to the outcome of the scenario with four routing agents (where all the pre-processing steps are performed concurrently), the running time is significantly lower. This has two reasons: On the one hand, no communication between agents is required. On the other hand, the hardware utilization is higher in the simulation, because the algorithm can answer four requests concurrently at any time. When a computation is finished, the next shortest path is computed immediately afterwards. In contrast, with four routing agents each agent computes its assigned shortest-path requests. When an agent has finished its computation, it has to wait until the last agent has finished its computations as well to ensure the correctness of the simulation which implements a *conservative synchronization mechanism* (cf. Sect. 2). If the agents would continue to perform tasks of the next time slice, the already processing agent might receive messages from the future and the consistency of the simulation is not ensured. The next time slice is only started when all agents have finished their tasks of the current time slice.

6   Conclusion and Outlook

Although it is also possible to increase the scalability of simulation platforms [3] or adding some optimization support for executing simulations on parallel processors [38], shortest-path searches remain one of the most cost-intensive operations of the agents in simulated logistic scenarios. In general, this is especially true for multiagent systems which are applied in real-world transport processes.

The results prove that standard search algorithms can preclude a system's industrial application. An efficient high speed shortest-path algorithm such as hub labeling on contraction hierarchies is a key component and essential for the industrial application of agent-based dispatching systems.

Moreover, the results reveal that a technical slight restriction of the autonomy of agents by applying a single algorithm saved in a static variable (which is part of all the agents) leads clearly to the lowest runtime of the simulation and to the lowest memory consumption.

As long as all agents run on the same machine (and on the same JVM), the disadvantage of less autonomy in this modeling approach is rather of a theoretical meaning than practically relevant. For instance, the robustness could even be guaranteed by a second redundant static instance of the algorithm. The privacy is also guaranteed, because the agents must not reveal their search queries to any other agent.

However, if the *full* autonomy of the agents has to be guaranteed, another option is to create several routing agents that receive routing requests, perform shortest-path searches, and provide the results. Although this approach consumes more time, i.e., because of the increased time consumption for message transfer and synchronization of agents, it can still profit from concurrent calculations as long as the number of routing agents is approximately equal to or lower than the number of available cores. Otherwise, the redundant algorithms consume a high amount of memory (in particular if the shortest-path searches are performed on large graphs) as well as time for communication and computation. In addition, shortest paths are not performed physically concurrently. In the extreme case it is even preferable that each consumer agent has its own algorithm. In this case the autonomy of the agents is maximized and less communication is required.

In conclusion, applying a static hub labeling algorithm in MAS for transport logistics, which is part of all agents, allows for concurrent calculations, improves the runtime performance of the simulation significantly, and reduces the memory usage. In contrast to the yet established modeling approaches, this allows the application of MAS in real-world logistic scenarios on large infrastructures with less hardware requirements.

Future research will focus on the application of shortest-path algorithms on several distributed machines. The PlaSMA simulation platform already supports the simulation of multiple agents that run on containers located on different machines. Moreover, we will investigate the behavior of shortest-path algorithms for dynamically changing graphs. In this case, the hub labels must be updated after the pre-processing is finished.

Acknowledgments. The presented research was funded by the German Research Foundation (DFG) within the project Autonomous Courier and Express Services (HE 989/14-1) at the University of Bremen, Germany.

References

1. Abraham, I., Delling, D., Goldberg, A.V., Werneck, R.F.: A hub-based labeling algorithm for shortest paths in road networks. In: Pardalos, P.M., Rebennack, S. (eds.) SEA 2011. LNCS, vol. 6630, pp. 230–241. Springer, Heidelberg (2011)
2. Abraham, I., Delling, D., Goldberg, A.V., Werneck, R.F.: Hierarchical hub labelings for shortest paths. In: Epstein, L., Ferragina, P. (eds.) ESA 2012. LNCS, vol. 7501, pp. 24–35. Springer, Heidelberg (2012)
3. Ahlbrecht, T., Dix, J., Köster, M., Kraus, P., Müller, J.P.: A scalable runtime platform for multiagent-based simulation. Technical report IfI-14-02, TU Clausthal (2014)
4. Barbucha, D., Jedrzejowicz, P.: Multi-agent platform for solving the dynamic vehicle routing problem. In: Proceedings of the Eleventh International IEEE Conference on Intelligent Transportation Systems, pp. 517–522 (2008)
5. Batz, G.V., Geisberger, R., Neubauer, S., Sanders, P.: Time-dependent contraction hierarchies and approximation. In: Festa, P. (ed.) SEA 2010. LNCS, vol. 6049, pp. 166–177. Springer, Heidelberg (2010)
6. Bauer, R., Columbus, T., Katz, B., Krug, M., Wagner, D.: Preprocessing speed-up techniques is hard. In: Calamoneri, T., Diaz, J. (eds.) CIAC 2010. LNCS, vol. 6078, pp. 359–370. Springer, Heidelberg (2010)
7. Bellifemine, F., Caire, G., Greenwood, D.: Developing Multi-Agent Systems with JADE. Wiley, Chichester (2007)
8. Bürckert, H.J., Fischer, K., Vierke, G.: Holonic transport scheduling with teletruck. Appl. Artif. Intell. 14(7), 697–725 (2000)
9. Christofides, N.: Worst-case analysis of a new heuristic for the travelling salesman problem. Technical report 388, Graduate School of Industrial Administration, Carnegie-Mellon University (1976)
10. Dijkstra, E.: A note on two problems in connexion with graphs. Numer. Math. 1(1), 269–271 (1959)
11. Dorer, K., Calisti, M.: An adaptive solution to dynamic transport optimization. In: Proceedings of the Fourth International Joint Conference on Autonomous and Multiagent Systems, AAMAS 2005, pp. 45–51. ACM, New York (2005)
12. Edelkamp, S., Gath, M.: Optimal decision making in agent-based autonomous groupage traffic. In: Filipe, J., Fred, A.L.N. (eds.) Proceedings of the Fifth International Conference on Agents and Artificial Intelligence (ICAART), vol. 1, pp. 248–254. SciTePress, Barcelona (2013)
13. Edelkamp, S., Gath, M.: Solving Single-vehicle pickup-and-delivery problems with time windows and capacity constraints using nested Monte-Carlo search. In: Duval, B., van den Herik, J., Loiseau, S., Filipe, J. (eds.) Proceedings of the Sixth International Conference on Agents and Artificial Intelligence (ICAART), vol. 1, pp. 22–33. SciTePress, Angers, France (2014)
14. Fischer, K., Müller, J.P., Pischel, M.: Cooperative transportation scheduling: an application domain for DAI. J. Appl. Artif. Intell. 10(1), 1–33 (1996)
15. Flood, M.M.: The traveling-salesman problem. Oper. Res. 4(1), 61–75 (1956)
16. Gamma, E., Johnson, E.R., Helm, R., Vlissides, J.: Design Patterns: Elements of Reusable Object-Oriented Software. Addison-Wesley, Reading (1995)
17. Gath, M., Herzog, O., Edelkamp, S.: Autonomous and flexible multiagent systems enhance transport logistics. In: 2014 11th International Conference Expo on Emerging Technologies for a Smarter World (CEWIT), pp. 1–6, October 2014

18. Gath, M., Edelkamp, S., Herzog, O.: Agent-based dispatching enables autonomous groupage traffic. J. Artif. Intell. Soft Comput. Res. (JAISCR) **3**(1), 27–42 (2013)
19. Gehrke, J.D., Schuldt, A., Werner, S.: Quality Criteria for multiagent-based simulations with conservative synchronisation. In: Rabe, M. (ed.) 13th ASIM Dedicated Conference on Simulation in Production and Logistics, pp. 545–554. Citeseer, Fraunhofer IRB Verlag, Stuttgart (2008)
20. Geisberger, R., Sanders, P., Schultes, D., Delling, D.: Contraction hierarchies: faster and simpler hierarchical routing in road networks. In: McGeoch, C.C. (ed.) WEA 2008. LNCS, vol. 5038, pp. 319–333. Springer, Heidelberg (2008)
21. Geisberger, R., Sanders, P., Schultes, D., Vetter, C.: Exact routing in large road networks using contraction hierarchies. Transp. Sci. **46**(3), 388–404 (2012)
22. Glaschenko, A., Ivaschenko, A., Rzevski, G., Skobelev, P.: Multi-agent real time scheduling system for taxi companies. In: Proceedings of the Eighth International Conference on Autonomous Agents and Multiagent Systems, AAMAS 2009, pp. 29–36 (2009)
23. Greulich, C., Edelkamp, S., Gath, M.: Agent-based multimodal transport planning in dynamic environments. In: Timm, I.J., Thimm, M. (eds.) KI 2013. LNCS, vol. 8077, pp. 74–85. Springer, Heidelberg (2013)
24. Greulich, C., Edelkamp, S., Gath, M., Warden, T., Humann, M., Herzog, O., Sitharam, T.G.: Enhanced shortest path computation for multiagent-based intermodal transport planning in dynamic environments. In: Filipe, J., Fred, A.L.N. (eds.) 5th International Conference on Agents and Artificial Intelligence (ICAART), vol. 2, pp. 324–329. SciTePress, Barcelona, 15–18 February 2013
25. Himoff, J., Skobelev, P., Wooldridge, M.: MAGENTA technology: multi-agent systems for industrial logistics. In: Proceedings of the Fourth International Joint Conference on Autonomous Agents and Multiagent Systems, AAMAS 2005, pp. 60–66. ACM, New York (2005)
26. Himoff, J., Rzevski, G., Skobelev, P.: Magenta technology multi-agent logistics i-scheduler for road transportation. In: Proceedings of the Fifth International Joint Conference on Autonomous Agents and Multiagent Systems, AAMAS 2006, pp. 1514–1521. ACM, New York (2006)
27. Jennings, N.R.: An agent-based approach for building complex software systems. Commun. ACM **44**(4), 35–41 (2001)
28. Jennings, N.R., Wooldridge, M.: Applications of Intelligent Agents. Springer-Verlag, New York (1998)
29. Kalina, P., Vokřínek, J.: Parallel solver for vehicle routing and pickup and delivery problems with time windows based on agent negotiation. In: Proceedings of the IEEE International Conference on Systems, Man, and Cybernetics (SMC), pp. 1558–1563 (2012)
30. Kohout, R., Erol, K.: In-time agent-based vehicle routing with a stochastic improvement heuristic. In: Proceedings of the 16th Conference on Artificial Intelligence and the 11th on Innovative Applications of Artificial Intelligence (AAAI/IAAI 1999), pp. 864–869. AAAI Press, Menlo Park (1999)
31. Leong, H.W., Liu, M.: A multi-agent algorithm for vehicle routing problem with time window. In: Proceedings of the 2006 ACM Symposium on Applied Computing, SAC 2006, pp. 106–111. ACM, New York (2006)
32. van Lon, R.R., Holvoet, T., Vanden Berghe, G., Wenseleers, T., Branke, J.: Evolutionary synthesis of multi-agent systems for dynamic dial-a-ride problems. In: Proceedings of the 14th Annual Conference Companion on Genetic and Evolutionary Computation, GECCO 2012, pp. 331–336. ACM, New York (2012)

33. MacQueen, J., et al.: Some methods for classification and analysis of multivariate observations. In: Proceedings of the Fifth Berkeley Symposium on Mathematical Statistics and Probability, vol. 1, pp. 281–297. California, USA (1967)
34. Mahr, T., Srour, J., de Weerdt, M., Zuidwijk, R.: Can agents easure up? A comparative study of an agent-based and on-line optimization approach for a Drayage problem with uncertainty. Transp. Res. Part C Emerg. Technol. **18**(1), 99–119 (2010)
35. Mes, M., van der Heijden, M., van Harten, A.: Comparison of agent-based scheduling to look-ahead heuristics for real-time transportation problems. Eur. J. Oper. Res. **181**(1), 59–75 (2007)
36. Müller, H.J.: Towards agent systems engineering. Data Knowl. Eng. **23**(3), 217–245 (1997)
37. Perugini, D., Lambert, D., Sterling, L., Pearce, A.: A distributed agent approach to global transportation scheduling. In: Proceedings of the IEEE/WIC International Conference on Intelligent Agent Technology (IAT 2003), pp. 18–24 (2003)
38. Sano, Y., Kadono, Y., Fukuta, N.: A performance optimization support framework for GPU-based traffic simulations with negotiating agents. In: Proceedings of 7th International Workshop on Agent-based Complex Automated Negotiations (ACAN2014) (2014)
39. Thangiah, S.R., Shmygelska, O., Mennell, W.: An agent architecture for vehicle routing problems. In: Proceedings of the 2001 ACM Symposium on Applied Computing, SAC 2001, pp. 517–521. ACM, New York (2001)
40. Vokřínek, J., Komenda, A., Pěchouček, M.: Agents towards vehicle routing problems. In: Proceedings of the Ninth International Conference on Autonomous Agents and Multiagent Systems, AAMAS 2010, vol. 1, pp. 773–780. International Foundation for Autonomous Agents and Multiagent Systems, Richland, SC (2010)
41. Warden, T., Porzel, R., Gehrke, J.D., Herzog, O., Langer, H., Malaka, R.: Towards ontology-based multiagent simulations: the PlaSMA approach. In: Bargiela, A., Azam Ali, S., Crowley, D., Kerckhoffs, E.J. (eds.) Proceedings of the European Conference on Modelling and Simulation, pp. 50–56. ECMS 2010 (2010)
42. Zhenggang, D., Linning, C., Li, Z.: Improved multi-agent system for the vehicle routing problem with time windows. Tsinghua Sci. Technol. **14**(3), 407–412 (2009)

SAJaS: Enabling JADE-Based Simulations

Henrique Lopes Cardoso[1,2]($\boxtimes$)

[1] Dep. Eng. Informática, Faculdade de Engenharia,
Universidade do Porto, Porto, Portugal
hlc@fe.up.pt
[2] LIACC – Laboratório de Inteligência Artificial e
Ciência de Computadores, Porto, Portugal

Abstract. Multi-agent systems (MAS) are widely acknowledged as an appropriate modelling paradigm for distributed and decentralized systems, where a (potentially large) number of agents interact in non-trivial ways. Such interactions are often modelled defining high-level interaction protocols. Open MAS typically benefit from a number of infrastructural components that enable agents to discover their peers at run-time. On the other hand, multi-agent-based simulations (MABS) focus on applying MAS to model complex social systems, typically involving a large agent population. Several MAS development frameworks exist, but they are often not appropriate for MABS; and several MABS frameworks exist, albeit sharing little with the former. While open agent-based applications benefit from adopting development and interaction standards, such as those proposed by FIPA, MABS frameworks typically do not support them. In this paper, a proposal to bridge the gap between MAS simulation and development is presented, including two components. The Simple API for JADE-based Simulations (SAJaS) enhances MABS frameworks with JADE-based features. While empowering MABS modellers with modelling concepts offered by JADE, SAJaS also promotes a quicker development of simulation models for JADE programmers. In fact, the same implementation can, with minor changes, be used as a large scale simulation or as a distributed JADE system. In its current version, SAJaS is used in tandem with the Repast simulation framework. The second component of our proposal consists of a MAS Simulation to Development (MASSim2Dev) tool, which allows the automatic conversion of a SAJaS-based simulation into a JADE MAS, and vice-versa. SAJaS provides, for certain kinds of applications, increased simulation performance. Validation tests demonstrate significant performance gains in using SAJaS with Repast when compared with JADE, and show that the usage of MASSim2Dev preserves the original functionality of the system.

Keywords: Multi-agent systems · Multi-agent based simulation · Model conversion · Standards

© Springer-Verlag Berlin Heidelberg 2015
N.T. Nguyen et al. (Eds.): TCCI XX, LNCS 9420, pp. 158–178, 2015.
DOI: 10.1007/978-3-319-27543-7_8

1   Introduction

The field of multi-agent systems (MAS) studies how to model complex systems using agents – autonomous, intelligent entities exhibiting social abilities that enable them to interact with each other [21]. Agent-based applications are in widespread use in multiple fields, both in research and industry. Such applications can be heterogeneous, often requiring interoperation between agents from different systems. In order to make this possible, agent technologies have matured and standards have emerged to support the interaction between agents.

The specifications of the Foundation for Intelligent Physical Agents (FIPA)[1] promote interoperability in heterogeneous agent systems. These standards define not only a common Agent Communication Language (ACL), but also a group of interaction protocols, recommended facilities for agent management and directory services [16].

Several frameworks exist [2,14] that offer some level of abstraction for a proper development of agent-based applications, allowing programmers to focus on a more conceptual approach in MAS design. However, only a few of them support FIPA standards (the most notable being JADE [4]), making interoperation between agents developed using different frameworks more difficult (although some claim to be interoperable with JADE, e.g. Jason [5] and SeSAm [12]).

Multi-agent based simulations (MABS) focus on applying MAS to model complex social systems, involving a large agent population. Simulations are sometimes used in the course of development of a full-featured MAS, for the purpose of testing. However, most platforms for MAS development are not well suited for MABS due to scalability limitations [13,17]. Popular agent-based simulation (ABS) frameworks, such as Repast [15] and NetLogo [19], lack support on advanced agent programming and multi-agent features, such as communication and infrastructure components. Given their social sciences background, it could be said that the kinds of agents such frameworks are best at modelling are not the same kind of agents considered in the multi-agent systems research community.

Still, agent-based simulation frameworks are widely used for MABS, given their support for large-scale simulations through the use of schedulers, environment spaces, data-collection and visualization facilities. In fact, there are potential gains in performance when running a MAS on top of a native simulation framework, which enables efficient large-scale testing of specific MAS properties.

Given this state of affairs, there is a growing interest in solutions that provide a richer set of programming tools for developing MABS, such as those typically available in MAS development frameworks. At the same time, an opportunity exists to partially automate the development of robust MAS from a previously tested simulation [12]. This would comprise a "write once, simulate and deploy" philosophy, where the same code could be used to run a large-scale simulation and to deploy the MAS in a distributed way.

---

[1] http://www.fipa.org/.

These two points are exactly the research directions taken in this paper. We focus on two popular frameworks for MAS development and simulation, respectively: JADE and Repast. These choices are related with the fact that both of these frameworks are open-source, have a wide and lively user community and extensive online support.

JADE [4] is a FIPA-compliant, general-purpose (i.e. not focused on a single domain) framework used in the development of distributed agent applications. It is a very popular MAS development framework that allows the creation of seamless distributed agent systems and complies with FIPA standards. It uses an architecture based on agent *containers* which allows the abstraction from the network layer, meaning that there is no difference, from the programmers perspective, between interactions among agents running in the same or separate machines. In terms of agent programming, JADE proposes the concept of *behaviour* as the building block for defining the tasks agents are able to execute. However, experiments with JADE show that the platform's scalability is limited [13]. Its multi-threaded architecture falls short in delivering the necessary performance to run a local simulation with a large number of agents, meaning that JADE is not an appropriate tool to create MABS.

Repast [6,15] is an agent-based modelling and simulation system that allows creating simulations using rich GUI elements and real time agent statistics. It can easily handle large numbers of agents in a single simulation. The former "flavour" of Repast – Repast 3 [6] – is still, for its higher simplicity, widely used, in particular by skilled Java programmers. The current Repast suite includes Repast Simphony [15], which comprises a significantly different approach to develop agent-based simulations, including visual tools for non-programmers, and ReLogo and Java APIs. Unlike JADE, though, Repast lacks much of the infrastructure for multi-agent management and interaction. Furthermore, programming agents in Repast is a task that starts from basic Java objects, without any conceptual support for agent development.

The main motivation for this work is thus to facilitate the development of rich multi-agent based simulations taking advantage of agent-based simulation frameworks. In the end, it should be straightforward to produce a simulation of a MAS more complex than those typically developed with such frameworks. Furthermore, code written for the simulation should be portable to the full-featured version of the underlying MAS.

In order to develop an integrated solution for bridging the domains of simulation and development of MAS, two main goals were pursued:

1. First, the creation of an adapter or API that allows MAS developers to abstract from simulation framework features and use familiar ones present in MAS development frameworks, thus creating "MAS-like MABS".
2. Second, the development of a MABS-MAS conversion tool. Having a MABS that is close to its underlying MAS makes it feasible and straightforward to engineer a tool that performs automatic conversion of MABS into equivalent MAS and vice-versa.

Given our choice for JADE, this solution is particularly useful for JADE developers who need to create a simulation of their already-developed MAS. By converting their code, the developer can run simulations and perform tests and fixes, later converting the simulation back to a MAS, preserving all changes. JADE developers can also create multi-agent simulations from scratch, using frameworks such as Repast, but taking advantage of familiar JADE-like features. Such simulations would then be converted to full-featured JADE MAS. Finally, the approach is also of interest to Repast developers who desire to expand their knowledge of MAS development using more complex frameworks. We are aware that this kind of facilities is only valuable for true multi-agent based simulation, and not in general for any agent-based modelling and simulation approach (for which Repast is primarily suited).

The rest of this paper is structured as follows. Section 2 presents related work, mainly devoted at bridging the gap between MAS simulation and development tools. Section 3 provides an overview of the whole solution proposed in this paper, presenting SAJaS and MASSim2Dev. Sections 4 and 5 describe the developed contributions – SAJaS and MASSim2Dev – in more detail, including their design choices and use cases. Section 6 explains how both tools have been validated. Section 7 presents some conclusions and Sect. 8 points lines of future work.

2   Related Work

Closing the gap between simulation and development of multi-agent systems has been identified as an important research direction. In a more comprehensive approach (as compared with the work reported in this paper), the fields of agent-oriented software engineering, on one hand, and agent-based modelling and simulation, on the other, can fruitfully be integrated, as suggested in the survey by Fortino and North [9].

The opportunity for applying multi-agent simulation in domains other than pure social simulation has been identified long ago. Davidsson [7] points out this trend by highlighting some properties that make MABS appealing to other domains, namely those requiring more sophisticated agent behaviours.

The idea of enriching agent-based simulation frameworks with multi-agent features is not new. In their work on SeSAm [12], Klügl et al. propose using agent-based modelling and simulation for software development. They focus on designing agent systems whose validity and robustness is ensured by prior verification through simulation.

Several frameworks exist that offer support to the development of MAS or MABS. Some are domain specific, meaning that their purpose was well defined in their conception: MASeRaTi [1] and MATSim [3] are some examples of MABS frameworks for traffic and transport simulations; PlaSMA [20] was designed for the logistics domain. Other frameworks like Repast [6,15], NetLogo [19] and GALATEA [8] are considered general-purpose. This enumeration is not meant to be exhaustive, including only a few examples of open-source tools.

A few attempts have been made in the direction of enriching simulation platforms with (multi-)agent features. Sakellariou et al. [18] proposed extending

NetLogo with BDI agent programming, as well as FIPA-ACL-like communication. Their choice of NetLogo is based on its potential use as an educational platform for simulation and multi-agent systems.

In the line of our own work, in the literature we can find a few proposals to bridge the gap between MAS development and simulation by integrating JADE with simulation features, either by extending this framework with a simulation layer created from scratch, or by integrating it with an existing simulation framework, such as Repast. Some of these are discussed in the next section.

2.1   JADE Simulation Extensions

MISIA [10] is a middleware whose goal is to enhance the simulation of intelligent agents and to allow the visualization and analysis of agent's behaviour. It is no longer an active project, having evolved into other more specific tools.

MISIA's approach consists of using a middle layer that acts as the bridge between two other layers that interact with JADE and Repast. By extending the agents in Repast and JADE, communicating through a coordinator and synchronizing their state, these agents work as a single one.

One of the challenges identified by the authors when re-implementing FIPA interaction protocols was synchronizing them with the Repast tick-based simulation model. Given JADE's event-driven architecture, MISIA proposes the use of a coordinator agent that informs the JADE-Agent when a tick has passed. It also proposes its own implementation of the interaction protocols supported by JADE, making them "tick-friendly".

JRep [11] proposes integrating JADE and Repast Simphony in a way that combines the macro and micro perspectives of the system with an interaction layer. JRep's approach is not as complex as MISIA's. By having the Repast Simphony agent encapsulate a JADE agent representation, synchronization is immediate and is assured without requiring an external coordinator. The two agent representations take care of synchronizing any state changes.

Each agent takes care of interfacing its respective framework. The interaction between agents in JRep is performed using FIPA ACL and the protocol implementations are those provided by the JADE platform. Similarly to MISIA, an Agent Representation Interface is used to introduce the concept of schedule in the JADE agent.

Unlike the two previous frameworks, the PlaSMA [20] system is based solely on the JADE platform. The distributed simulation is synchronized by entities called "Controllers" who communicate with the "Top Controller", keeping the pace of the simulation and handling agent lifecycle management as well. Unlike MISIA and JRep, PlaSMA is still an active project.

2.2   Limitations

Distributed simulation of multi-agent systems brings non-desirable scalability issues, mainly due to synchronization overheads [17]. Furthermore, scenarios with a high communication-to-computation ratio [13], which are typical in many multi-agent applications, are largely affected by network connectivity, bringing large communication overheads. If it is often the case that agents remain idle until a message is received, no real advantage is attained from having a distributed simulation. For this reason, general-purpose multi-agent platforms with multi-threading support, while being useful for deploying MABS on distributed resources, are not a viable approach for large-scale simulations.

JADE is a rich and powerful platform. As pointed out before, however, for many multi-agent simulation scenarios its overhead has a significant impact on simulation performance [13]. Repast Simphony is a very efficient simulation platform. However, it lacks support for agent programming concepts and multi-agent features, such as high-level communication and infrastructural components.

Even though both MISIA and JRep attempt to integrate the best of JADE and Repast, they still rely on JADE to run the agents themselves. As far as Repast simulations are concerned, JADE's multi-threaded infrastructure affects performance very significantly. This would be the main drawback of these approaches. The same is true for PlaSMA, naturally.

As we will describe in the following sections, the distinguishing feature of our approach is the possibility of using Repast with JADE features, without the need to interface with the JADE runtime system. In order to do that, we replace the JADE Agent class with a Repast-friendly one, whose scheduled execution we are able to control directly.

3   Bridging JADE-Based Simulation and Development

As mentioned in Sect. 1, this work aims at enabling the development of rich multi-agent based simulations, capturing features available in a MAS development framework – JADE in our case. Besides taking advantage of a simulation infrastructure (such as Repast) to run the simulation proper, we also aim at using the same developed code both for the simulated run and for the actual deployment of the MAS (an approach that has been also suggested in [12]).

The contributions reported in this paper are two-fold:

- The **Simple API for JADE-based Simulations (SAJaS)** is an adapter API that enables running JADE-like simulations, connecting the underlying MAS with a simulation framework. Our rationale was to be as surgical as possible, seeking to take advantage of most JADE features which do not directly affect simulation performance.
- The **MAS Simulation to Development (MASSim2Dev)** code conversion tool is an Eclipse plug-in that offers a seamless automatic conversion between JADE and SAJaS (and vice versa). In order to do that, a mapping between JADE classes and their equivalent in SAJaS is provided, making the tool mostly independent of those two APIs.

SAJaS is an API meant to be used with simulation frameworks, enriching them with JADE-based features, such as behaviour-based agent programming, interaction protocols and agent management services. For this reason, only runtime-specific JADE classes have been replaced by new versions that enable the simulation framework to take control over agent execution. SAJaS own classes are very similar to their JADE counterparts, in order to facilitate code conversion with MASSim2Dev. More importantly, this allows proficient JADE developers to create SAJaS-based simulations using a familiar JADE-based API.

SAJaS was initially created to be used with Repast Simphony. However, SAJAs design choices consider its straightforward integration with other simulation frameworks, as we will illustrate later.

MASSim2Dev currently provides programmers with two possible actions: (i) given a JADE-based project, convert it to a SAJaS-based project; (ii) given a SAJaS-based project, convert it to a JADE-based project.

3.1 FIPA Specifications

The need to extend simulation frameworks with multi-agent features is best addressed if we take into consideration agent technology standards, such as those proposed by FIPA. Since JADE is a FIPA-compliant platform, basing our approach on its implementation of FIPA standards allows us to inherit these multi-agent systems development features.

Through JADE, SAJaS includes FIPA standards divided into two broad categories: Agent Management and Agent Communication.

FIPA Agent Management specifications include the Directory Facilitator (DF) and the Agent Management Service (AMS). The DF is a component that provides a yellow page service. It endows agents with run-time register and search facilities that enables agents to announce themselves to the rest of the MAS and to find out about other agents in the system. The AMS is meant to manage the agent platform, namely creating and terminating agents. Agent registration in the AMS is mandatory, and results in the assignment of an agent identified (AID), needed e.g. for communication purposes. Communication is supported by a Message Transport System (MTS).

FIPA Agent Communication specifications include the notions of ACL Message, Communicative Acts and Interaction Protocols. An ACL message includes an "envelope" that contains several fields with communication details. Exploiting those fields, message templates may be used to filter incoming messages, allowing an agent to process them selectively. A communicative act is a central part of an ACL message, and is meant to disclose the communicative intention of the sender. FIPA Interaction Protocols typify communication interactions among agents by specifying two roles: initiator (the agent starting the interaction) and responder (a participant in the interaction). Each protocol defines precisely which messages are sent by each role and in which sequence.

3.2 JADE and Repast

Given our choices on JADE and Repast as target frameworks, we need to properly understand the main features and differences among them. As Table 1 shows,

JADE agents execute in separate threads, which enables distributing agents among different machines. The downsize of this approach is its impact on performance when running locally a significant number of agents, which is a typical scenario in simulation. In fact, experiments with JADE have shown that the platform's scalability is limited: the global system performance drops quickly for large numbers of agents [13]. This further strengthens the idea that using JADE or a JADE-Repast hybrid, as described in Sect. 2.1, is not the best course of action if performance is an important issue.

In JADE, agents are distributed across *containers*. Each host machine can contain multiple containers, and JADE allows agents in different containers and hosts to interact with each other through message exchange. In each JADE instance, a main container exists where some special agents reside (namely, the Agent Management System and the Directory Facilitator), which help in the management and address resolution of the agents. JADE agents can even hop into another container.

While not having an equivalent infrastructure, Repast Simphony does include the notion of *context* that indexes all scheduled objects. Because Repast does not support distributed applications, there is no need for address resolution services.

JADE agent actions can be executed during setup and takedown, but most are encapsulated in objects called *behaviours*. JADE has many different kinds of behaviours that function in different ways, such as running one single task once or running cyclically. Other behaviours implement FIPA interaction protocols, which agents can use to interact with other agents. Together with ontology support, this comprises one of the most useful features of JADE when developing MAS. Given this strong support for communication, agents can be said to execute in an event-driven way in scenarios that rely strongly on interaction among them.

In Repast Simphony, agent execution is scheduled explicitly. Any class added to Repast's context can contain Java annotations that indicate which methods should be called and when (for instance, in every simulation tick or from time to time). Alternatively, an explicit *scheduler* may be used to define which actions to execute throughout simulation. While this approach is very flexible, more complex structures supporting agent development are non-existent in Repast. There is also no support to agent communication, which must be programmed through direct method invocations.

3.3   Usage Scenarios

The rationale behind the design of SAJaS and MASSim2Dev foresees a number of possible usage scenarios. Figure 1 illustrates the scenarios where this system is expected to be useful.

One possible scenario concerns a JADE developer who wishes to perform some tests and simulations of his JADE-based MAS, by running the system in a local and controlled environment. The developer can use MASSim2Dev to convert the MAS into a SAJaS MABS. Eventually, the application can be converted back if changes were introduced while performing tests.

Table 1. Comparison of JADE and Repast features.

	JADE	Repast
Distributed	Yes	No
Simulation tools	No	Yes
Scalability	Limited	High
Open source	Yes	Yes
Agent execution	Behaviours	Scheduler
	Multi-thread	Single-thread
	Event-driven	Tick-driven
	Asynchronous	Synchronous
Interaction	FIPA ACL	Method calls
		Shared resources
Ontologies	Yes	No

A second possible scenario could be one where a developer intends to create a MABS with the goal of later converting it to a full-featured MAS. The developer could be fluent in Repast, desiring to create agent simulations that take advantage of communication and agent management tools (present in JADE and SAJaS); the developed could also be experienced in JADE, intending to create Repast simulations using familiar JADE-like tools.

A third scenario may consist of a researcher that simply wants to create a complex agent-based, FIPA-compliant simulation. In this case, there is no need for a code conversion tool, but SAJaS can be used as a standalone library.

The next sections describe in detail both the SAJaS API and the MAS-Sim2Dev conversion tool.

4   SAJaS

As its name implies, the Simple API for JADE-based Simulations has been design by taking the most advantage of MAS development features offered by JADE, with the aim of providing a simulation development experience that is comfortable for JADE experienced programmers. From the point of view of the MAS programmer, working with the SAJaS API feels the same as working with JADE, although with some (minor) limitations, as we will later explain in Sect. 4.2. To achieve this result, only the components related with the JADE runtime infrastructure have been replaced to enable the simulation framework to control agent execution. Even so, the few core SAJaS classes are API-wise equivalent to their JADE counterparts. This facilitates usage from the point of view of the programmer and conversion through MASSim2Dev.

The most evident feature that is not supported in SAJaS is JADE's network layer that enables the creation of distributed MAS. This is, in fact, the one feature that we consider to negatively affect communication-intensive large-scale simulation performance.

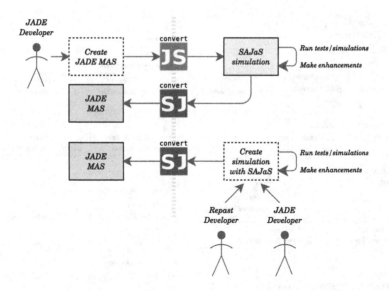

Fig. 1. Possible work flows for SAJaS/MASSim2Dev users ("SJ" and "JS" represent conversion from SAJaS to JADE and the reverse, respectively).

Figure 2 shows a basic class diagram of SAJaS, where for clarity not all dependencies are visible. Apart from their package distribution (which follows JADE very closely), SAJaS classes can be grouped as follows:

- *Core classes*: The need to reimplement the `Agent` class arose from the fact that in SAJaS each agent is no longer an independent thread. We thus need control over each agent's execution. Apart from that, most of JADE's own implementation of this class was kept. The reason for not extending `jade.core.Agent` is due to non-public method declarations that we needed to override. The new `AID` class, which extends the JADE version, provides a means to properly set and use a platform name.
- *Runtime infrastructure*: These classes have the purpose of launching the system, creating containers and agents, and starting agent execution: `Runtime`, `PlatformController`, `ContainerController` and `AgentController`.
- *FIPA services*: Classes `FIPAService`, `AMSService`, `DFService` and `DFAgent` correspond to FIPA services available in any JADE MAS. In particular, `DFAgent` is our implementation of the yellow page service agent, which naturally extends `sajas.core.Agent`. FIPA services had to be reimplemented due to the blocking approaches available in the JADE API, which do not work in SAJaS given its single thread nature (more on this in Sect. 4.2).
- *Agent dependencies*: As a consequence of having a new `Agent` class, every class in the JADE API that makes use of it had to be included in SAJaS as well, simply to redirect references to the SAJaS `Agent` class – this includes behaviour (`sajas.core.behaviours`) and protocol (`sajas.proto`) classes. Naturally, the same redirection was needed in the previous groups of classes.

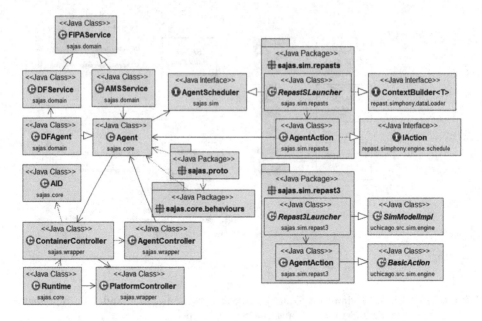

Fig. 2. A simplified UML class diagram of SAJaS.

– *Simulation interface*: This group of classes is specific of SAJaS, and represents its connection with the simulation infrastructure. The `AgentScheduler` interface allows agents to be added to the simulation scheduler; the specific scheduler to use is set statically in the `Agent` class. This approach makes it easy to integrate SAJaS with different simulation frameworks. Currently, two of them are included: Repast 3 and Repast Simphony. For the latter, classes `RepastSLauncher` and `AgentAction` implement the needed Repast Simphony interfaces (also shown in Fig. 2): `RepastSLauncher` is responsible for building the simulation context and setting the agent scheduler, after which application-specific JADE-related code is invoked (an abstract method is declared for that purpose); `AgentAction` implements the actual execution of scheduled agents. A similar pair of classes exists for Repast 3 (whose dependencies are omitted in Fig. 2).

4.1   Agent Execution and Interaction

JADE execution is concurrent and parallel, since JADE supports multi-threaded and distributed agent systems. Agent tasks are encapsulated in behaviours, which are executed in a sequential round-robin fashion. Multiple agents can be executing their behaviours simultaneously. It is up to the programmer to ensure that the application does not rely on the actual order of execution.

Execution in simulation frameworks like Repast, on the other hand, is not concurrent. Repast uses a time-share type of execution, granting each agent, in sequence but in no particular order, the right to perform its tasks. Scheduled methods (e.g. agent executions) typically run consecutively but with variable execution order. Again, it is up to the simulation designer to ensure that simulation results do not depend on the order of execution. We thus have a different scheduling granularity: while in JADE each agent schedules its own behaviours, in Repast it is the agents that are scheduled in a shared execution thread.

Given the lack of support in Repast for agent communication, the typical way of implementing this kind of agent interaction is through method calls or shared resources. Albeit feasible, this approach brings additional challenges when considering the implementation of complex interaction protocols, such as the risk of stagnation if agents engage in a "long conversation". By taking advantage of JADE features, however, we are able to maintain its asynchronous communication mode. This enables the use of conceptually concurrent interactions among agents. In JADE (and in SAJaS), each agent has a message queue to which messages are delivered by the messaging service, and processes such messages by consuming them in appropriate behaviours.

Looking from a different perspective, we can also say that while simulations in Repast usually depend on the synchrony of the environment, by using message-waiting behaviours we are able to maintain a synchronous execution, while simulating an asynchronous one. With this approach, we can easily define protocol-based milestones that can be exploited in the course of a simulation.

To better demonstrate the differences between agent execution in both frameworks, Figs. 3 and 4 represent a scenario where two agents send a message to a third one, who replies. In SAJaS single-threaded execution (Fig. 3), messages are delivered to agent C's message queue, and are processed only when it is C's turn to execute in the shared thread[2]. In JADE (Fig. 4), messages can arrive concurrently. Their arrival triggers an event and they are processed right away in the receiving agent's thread. In this case, agent C handles the messages as they arrive and issues the respective replies.

4.2 Current Limitations

As mentioned before, SAJaS takes a near-full advantage of JADE's features. The following are a couple of exceptions regarding the current version of SAJaS.

Handling Time. The FIPA ACL message structure specification includes an optional "reply-by" parameter, to be filled-in with the latest time by which the sending agent would like to receive a reply. This parameter may be of particular use in interaction protocols, by halting waiting for the next sequential message when the indicated time has elapsed. Given the simulation bias of SAJaS, it is not clear yet to which time this should refer to. Accelerating simulation execution means that we should not use reply-by values larger than strictly necessary, which typically depends on the application in mind. Translating such

---

[2] This scenario is merely hypothetical; relevant is the variable agent execution order.

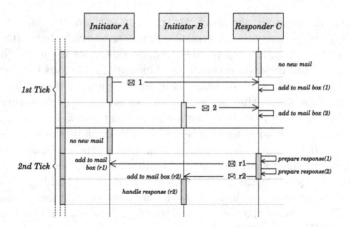

Fig. 3. Communication in SAJaS, in a shared execution thread.

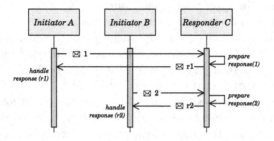

Fig. 4. Communication in JADE, each agent running in parallel.

timestamps to simulation ticks is probably the way to go, but enough simulation ticks should be allowed for responding agents to process the message and respond accordingly. This, in turn, requires having a mechanism that interfaces properly with simulation scheduling.

Two specific JADE behaviours are also hindered by the transition from JADE to SAJaS, and should thus be avoided. A `WakerBehaviour` specifies a task to be executed after a specific amount of time has elapsed. A `TickerBehaviour` specifies a periodic task to be executed in regular time intervals.

Blocking Approaches. Although discouraged by JADE, programmers may use so-called blocking approaches when interacting with each other. The effect of these approaches is that certain methods in JADE's API will only return after a message is received by the agent (or even after a complete protocol has terminated). This is achieved through a `blockingReceive` operation defined in JADE's `Agent` class, or indirectly by making use of `doFipaRequestClient` in the `FIPAService` class. Given the single-threaded approach of SAJaS, or more precisely of the simulation frameworks (such as Repast) it may be aligned with, it comes to no surprise that blocking approaches do not work. Another typical usage of blocking approaches concerns (de)registering and searching the DF

(JADE's yellow page service), for which a number of blocking methods are available in the `DFService` class. Although the same effects may be obtained using non-blocking approaches (by establishing a communication with the DF agent), in SAJaS we opted to reimplement the available blocking methods in `DFService`, making sure they do not block, while achieving the desired functionality.

5   MASSim2Dev

There are multiple ways to tackle the problem of code transformations. The brute force approach would be to parse the source code, create an abstract syntax tree (AST) which represents all code constructions in a program, perform certain transformations in the tree, and finally generate back the code from the new AST. Fortunately, there are free and open source projects that developers can use to do exactly this with significantly reduced effort.

The Eclipse Java Development Tools (JDT)[3] used to develop MASSim2Dev is a group of tools integrated in the Eclipse IDE. Some of its most interesting features include automatic project cloning, handling of classes, imports, methods and fields as objects, as well as the possibility of doing complex manipulation tasks without parsing the code. It does, however, allow the use of a high level AST for a more direct manipulation of the source code. JDT is accessible to plugin developers from within Eclipse.

MASSim2Dev is an Eclipse plugin that makes use of SAJaS. It acts as a translator that changes the MAS application/simulation dependencies on one framework (JADE/SAJaS) to the equivalent classes in the other framework. When converting, a new Eclipse Java project is created: if this is a JADE project (a conversion from SAJaS to JADE), references to SAJaS classes are redirected to their JADE equivalent. On the other hand, if the new project is a SAJaS-based one (a conversion from JADE to SAJaS), references to JADE classes that have been reimplemented in SAJaS are redirected to these new versions. Any other references to JADE's API are kept.

5.1   Plugin Execution

In its current version, MASSim2Dev simply includes a couple of buttons, activating the conversion of a JADE project to SAJaS, or of a SAJaS-based simulation to JADE, respectively. When one of such buttons is pressed, the plugin is activated, performing a sequence of actions. In the case of a SAJaS-to-JADE conversion, the plugin:

1. Clones the selected project;
2. Changes all references to SAJaS classes into their JADE equivalent;
3. Removes the no longer needed SAJaS library from the new project;
4. Fixes hierarchies (e.g. classes that extended `sajas.core.Agent` must now extend `jade.core.Agent`).

---

[3] https://www.eclipse.org/jdt/.

A similar sequence of actions is fired in the case of a JADE-to-SAJaS conversion. In that case, the SAJaS library is added do the project's build path.

In order to map class imports between JADE and SAJaS, a dictionary file is included. This approach accommodates future SAJaS upgrades, or its interface with further simulation frameworks, without having to change MASSim2Dev.

5.2 Handling JADE Updates

As mentioned in Sect. 4, when developing SAJaS we have tried to make the smallest possible changes to JADE's API, with the aim of incorporating in SAJaS-based simulations all the features that JADE programmers have available. Given the continuous development of JADE, however, new releases of that framework could imply a significant recoding of SAJaS.

It turns out that because of the generic mode of operation of MASSim2Dev, we are able to use it to perform most of that recoding effort, by providing a dictionary file that indicates which are the classes that need to be mapped. It should be evident from Sect. 4 which are the classes requiring our attention.

6 Validation

In order to illustrate the validation of both SAJaS and MASSim2Dev, in this paper we make use of two experimental scenarios that cover all relevant features. All experiments have been run[4] on three frameworks: JADE, SAJaS making use of Repast 3, and SAJaS making use of Repast Simphony.

The first experimental scenario covers most JADE programming features, and is described in Sect. 6.1, together with experimental runs and results.

The second experimental scenario starts from a JADE-based implementation of the Risk board game, which was developed prior to the start of this project. The Risk application is converted to SAJaS using MASSim2Dev. Section 6.2 describes this experimental scenario, together with experimental runs and results.

6.1 The Service Consumer/Provider Scenario

In this scenario, service consumers establish contract-net negotiations with service providers. A protocol initiator (the consumer) starts a FIPA-CONTRACT-NET by issuing a call-for-proposals (CFP) to all providers registered in the DF. Each responder (provider) PROPOSEs a price. Finally, the consumer chooses the cheapest proposal and replies with ACCEPT/REJECT-PROPOSALs accordingly. The execution of the service by the winning service provider may succeed or fail, as there are good and bad service providers. When the provider sees that it is going to fail, it may subcontract the service execution to another service provider randomly chosen, by sending a REQUEST (thus initiating a FIPA-REQUEST protocol); the subcontracted service provider may, again, succeed

---

[4] We have used a 64 bit Intel Core(TM)2 Duo CPU E8500, 3.16 GHz, 6 GB RAM machine.

or fail. An INFORM or a FAILURE message is sent to the service consumer, respectively. Some service consumers will be paying attention to service execution outcomes. In this case, they will start contract-net protocols only with a number of the best providers in the market, according to their own experience.

This scenario exploits most features that a programmer may want to make use of in JADE[5], including:

- The yellow page service, used for registering, searching and subscribing.
- Several kinds of behaviours, including protocol-related ones, cyclic behaviours and wrapper behaviours.
- Several kinds of protocols available in JADE, including FIPA-SUBSCRIBE (to the yellow page service), FIPA-CONTRACT-NET and FIPA-REQUEST, as well as responder dispatchers and register behaviour handlers.
- Languages and ontologies, used for the content of ACL messages.

We have run two sets of experiments based on this scenario. The first one tries to show the similarity of results when using each of the three frameworks (JADE, SAJaS+Repast3 and SAJaS+RepastS). While the exact scenario details are not determinant, we simply want to show that service consumers that filter out bad service providers tend to get more successfully executed services. In this experiment, we have 10 consumers negotiating with all providers, 10 consumers negotiating only with the five best providers, and 20 providers (half good, half bad). Each service provider makes random proposals within a fixed random range; furthermore, good providers have a 0.8 probability of successfully executing the service, while for bad providers this probability drops to 0.2. In each experiment, each consumer establishes a sequence of 100 contract net negotiations.

Figure 5 shows that the outcomes of simulation are similar for each of the three frameworks. Values shown comprise average results from 5 simulation runs. This proves that scenario conversion between JADE and SAJaS obtains equivalent systems. In fact, experiments free from random factors (which are not included in this paper) show identical results.

The second set of experiments compares execution times, for different numbers of agents. Figure 6 plots the results for each framework. The value of n represents the number of consumers of each type, while the total number of providers is $5 \times n$. Every thing else is configured as in the previous experiment.

It is clear that SAJaS, both when paired with Repast 3 or Repast Simphony, outperforms JADE in terms of simulation performance.

6.2   The Risk Board Game Scenario

RISK is a multi-player strategy board game played in turns[6]. The game implementation used for this experiment was developed with JADE before the conception of the project described in this paper. The game is played automatically

---

[5] We point the reader to the JADE documentation for details on these features.

[6] The reader can find details about Risk at http://en.wikipedia.org/wiki/Risk_(game).

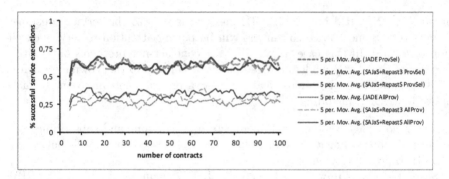

Fig. 5. Outcome comparison: thinner lines correspond to consumers negotiating with all providers (AllProv), while thicker lines correspond to those negotiating with the best providers only (ProvSel).

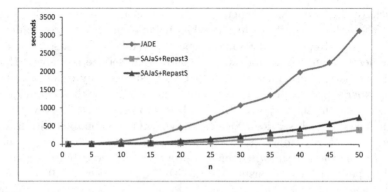

Fig. 6. Simulation performance for different numbers of agents (n consumers of each type and $5 \times n$ providers).

by software agents, competing against each other for the conquest of a map that loosely resembles a world map and its regions.

Playing agents have different playing skills and are classified as aggressive, defensive, opportunistic or random. Communication occurs between the players and the game agent using the FIPA-REQUEST protocol. The game also heavily relies on custom Finite State Machine Behaviours (supported by JADE through the FSMBehaviour class) to control game progress. To evaluate the performance of the game, logging features were introduced to the original source code of the application, in order to record the number of rounds executing in each second. No other changes were made to the original code.

For this experiment, a match with five "random agents" was set up. Random agents do not follow any particular strategy of attack, defence or soldier distribution; a game with random agents only is always never-ending. To analyse performance using different runtime frameworks, the game was converted from JADE to SAJaS using MASSim2Dev.

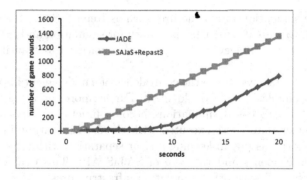

Fig. 7. Performance of a Risk match with 5 random agents.

The game was repeated 3 times. Average results are shown in Fig. 7, for the first 20 s of the game. As can be seen, SAJaS has a much better execution performance in an initial simulation period, which we attribute to the much faster setup phase as compared to JADE. Given the low number of agents that are executing and the turn-based nature of the Risk game, the two frameworks have a comparable execution performance after this first period. Although not imposing a strong overhead in terms of parallel execution threads, the executing behaviour of agents in Risk strongly relies on communication (agents are idle most of the time), which explains the fact that the lines in Fig. 7 are mostly parallel from second 15 onwards.

7 Conclusions

When developing multi-agent systems, it is useful to run simulations for the purpose of testing. Most MAS development frameworks, however, are not well suited for simulation, mainly due to scalability limitations. In fact, some of such frameworks, such as JADE, focus instead on deployment features of the developed MAS, such as the possibility to run the system distributed among machines – a crucial aspect if multi-agent systems are to be applied in real-world scenarios.

Agent-based simulation frameworks, on the other hand, do not offer significant support for agent and multi-agent programming, such as high-level communication and agent building blocks, leaving the programmer with simple Java objects to program with. Using simulation frameworks for simulating preliminary versions of multi-agent systems is thus a hassle, also because of the need to recode a significant part of the simulation if it is to be later deployed as a full-featured MAS.

Given the growing interest in solutions that provide a richer set of programming tools for developing MABS, our proposal is meant to take advantage of the best of both worlds, by providing a simulation engine-based infrastructure for developing multi-agent systems. Two advantages are offered with SAJaS. First, the MABS programmer has a rich set of multi-agent programming features offered by JADE, while being able to explore simulation-related features offered by the simulation infrastructure, such as Repast. Second, a same implementation can to a great extent be used both for simulation and for deployment purposes. The MASSim2Dev tool helps on automating this transition, in both directions.

Our experiments have shown the equivalence of a multi-agent implementation when used for simulation and deployment. Furthermore, by using a simulation infrastructure, JADE-based simulations become practical. Repast simulations scale much better than those executed with the JADE runtime infrastructure. A conscious effort was put in keeping a clear separation within SAJaS between Repast-specific elements and the core of SAJaS API. This enables the future integration of SAJaS with other simulation infrastructures.

It should be noted that simulation performance gains are not universal. The application scenarios where MAS simulations are expected to obtain higher gains by making use of SAJaS are those where agent interaction through communication plays a main role. If, on the other hand, communication overheads are bearable and outbalanced by the benefits of having a distributed simulation infrastructure that enables a parallel execution of the agents in the MAS, then SAJaS might not be the best approach. In such cases we can say that computation plays a more important role, and thus distributing it through a number of cores is a more sensible approach.

8   Future Work

Some lines of future work on SAJaS and MASSim2Dev will be pursuit. Although SAJaS is already strongly integrated with JADE, as mentioned in Sect. 4.2 it currently has at least two limitations. The notion of time is the most critical in terms of correspondence between the simulation and deployment versions of a multi-agent system. The fact that time is meaningful in a MAS is related to potential network problems or computation time. These issues mostly disappear when running a local simulation: no network problems can affect it, and in many cases agent-based simulations assume simple agent behaviours, with minimal computation times. It is therefore not clear how time-handling in JADE-based communication should be ported to SAJaS, if not to simply ignore the possibility of timeouts. This is something we intend to investigate.

The second limitation is related with JADE's blocking communication approaches. Encompassing this feature in SAJaS may be justified for extending the coverage of JADE features. But in any case, it is always possible to reprogram a MAS that makes use of these approaches to a version relying only on non-blocking approaches.

Apart from the possibility of exploiting message-waiting behaviours to achieve synchronisation, SAJaS does not include any other synchronization mechanism. This is an important distinction as compared to other works, such as [10]. Given its reliance on the underlying simulation framework, however, it should not be too difficult to implement a more robust synchronization approach, by upgrading the `sajas.core.Agent` class with appropriate data members and methods.

The modularity of SAJaS allows future extensions without changing the API and opens doors to future integration with simulation frameworks other than Repast. Doing so may enlarge the community of SAJaS potential users.

One interesting feature in Repast is the ability to create real time visualizations of simulation data. This is possible in part because agents in Repast are executed locally, so access to this data is facilitated[7]. It could be interesting to include data collection and display tools that could be ported between frameworks, taking advantage of MASSim2Dev.

Possible enhancements to the MASSim2Dev plugin include providing support for user configurations, such as the selection of the name and location of the newly generated project, and the automatic creation of "stub launchers" that would allow to quickly test if the generated project executes correctly. As mentioned in Sect. 5.2, with proper configuration MASSim2Dev can also be used to automatically generate new SAJaS versions triggered by JADE updates.

Finally, SAJaS and MASSim2Dev are being released[8] to the academic community for further development, discussion and use.

Acknowledgments. The author would like to thank João Lopes for his initial work on SAJaS and MASSim2Dev, and also João Gonçalves and Pedro Costa for providing the source code of their JADE-based Risk game implementation.

References

1. Ahlbrecht, T., Dix, J., Köster, M., Kraus, P., Müller, J.P.: A scalable runtime platform for multiagent-based simulation. Technical Report IfI-14-02, TU Clausthal (2014)
2. Allan, R.: Survey of Agent Based Modelling and Simulation Tools. Technical Report DL-TR-2010-007, Science and Technology Facilities Council, Warrington, U.K. (2010)
3. Balmer, M., Meister, K., Rieser, M., Nagel, K., Axhausen, K.W.: Agent-based simulation of travel demand: structure and computational performance of MATSim-T. ETH, Eidgenössische Technische Hochschule Zürich, IVT Institut für Verkehrsplanung und Transportsysteme (2008)
4. Bellifemine, F.L., Caire, G., Greenwood, D.: Developing multi-agent systems with JADE, vol. 7. Wiley, Chichester (2007)
5. Bordini, R.H., Hübner, J.F., Wooldridge, M.: Programming Multi-Agent Systems in AgentSpeak using Jason. Wiley, Chichester (2007)
6. Collier, N.: Repast: an extensible framework for agent simulation. The University of Chicagos Social Science Research **36** (2003)
7. Davidsson, P.: Multi agent based simulation: beyond social simulation. In: Moss, S., Davidsson, P. (eds.) MABS 2000. LNCS (LNAI), vol. 1979, pp. 97–107. Springer, Heidelberg (2001)

---

[7] We did not take advantage of these features when collecting results from our experiments, since they are not available in JADE.

[8] http://web.fe.up.pt/~hlc/doku.php?id=SAJaS.

8. Dávila, J., Uzcátegui, M.: Galatea: a multi-agent simulation platform. In: Proceedings of the International Conference on Modeling, Simulation and Neural Networks (2000)
9. Fortino, G., North, M.J.: Simulation-based development and validation of multi-agent systems. J. Simul. **7**(3), 137–143 (2013)
10. García, E., Rodríguez, S., Martín, B., Zato, C., Pérez, B.: MISIA: middleware infrastructure to simulate intelligent agents. In: Abraham, A., Corchado, J.M., González, S.R., De Paz Santana, J.F. (eds.) International Symposium on Distributed Computing and Artificial Intelligence. AISC, vol. 91, pp. 107–116. Springer, Heidelberg (2011)
11. Gormer, J., Homoceanu, G., Mumme, C., Huhn, M., Muller, J.: Jrep: extending repast simphony for jade agent behavior components. In: Proceedings 2011 IEEE/WIC/ACM International Conference on Web Intelligence and Intelligent Agent Technology, vol. 02, pp. 149–154. IEEE Computer Society (2011)
12. Klügl, F., Herrler, R., Oechslein, C.: From simulated to real environments: how to use SeSAm for software development. In: Schillo, M., Klusch, M., Müller, J., Tianfield, H. (eds.) MATES 2003. LNCS (LNAI), vol. 2831, pp. 13–24. Springer, Heidelberg (2003)
13. Mengistu, D., Troger, P., Lundberg, L., Davidsson, P.: Scalability in distributed multi-agent based simulations: the jade case. In: 2nd International Conference on Future Generation Communication and Networking Symposia (FGCNS'08), vol. 5, pp. 93–99. IEEE (2008)
14. Nikolai, C., Madey, G.: Tools of the trade: a survey of various agent based modeling platforms. J. Artif. Soc. Soc. Simul. **12**(2) (2009)
15. North, M., Howe, T., Collier N., Vos, R.: The repast simphony runtime system. In: Proceedings of the Agent 2005 Conference on Generative Social Processes, Models, and Mechanisms (2005)
16. O'Brien, P., Nicol, R.: Fipatowards a standard for software agents. BT Technol. J. **16**(3), 51–59 (1998)
17. Pawlaszczyk, D., Strassburger, S.: Scalability in distributed simulations of agent-based models. In: Winter Simulation Conference, WSC 2009, pp. 1189–1200 (2009)
18. Sakellariou, I., Kefalas, P., Stamatopoulou, I.: Enhancing netlogo to simulate BDI communicating agents. In: Darzentas, J., Vouros, G.A., Vosinakis, S., Arnellos, A. (eds.) SETN 2008. LNCS (LNAI), vol. 5138, pp. 263–275. Springer, Heidelberg (2008)
19. Tisue, S., Wilensky, U.: Netlogo: a simple environment for modeling complexity. In: International Conference on Complex Systems, pp. 16–21 (2004)
20. Warden, T., Porzel, R., Gehrke, J.D., Herzog, O., Langer, H., Malaka, R.: Towards ontology-based multiagent simulations: the plasma approach. In: European Conference on Modelling and Simulation, ECMS 2010, pp. 50–56, Kuala Lumpur, Malaysia, 1–4 June 2010
21. Wooldridge, M.: An Introduction to MultiAgent Systems, 2nd edn. Wiley Publishing, Chichester (2009)

Strategic Negotiation and Trust in Diplomacy – The DipBlue Approach

André Ferreira[1], Henrique Lopes Cardoso[1,2(✉)], and Luís Paulo Reis[2,3]

[1] Dep. Eng. Informática, Faculdade de Engenharia,
Universidade do Porto, Porto, Portugal
andre.ferreira.v2@gmail.com, hlc@fe.up.pt
[2] LIACC – Laboratório de Inteligência Artificial
e Ciência de Computadores, Porto, Portugal
[3] DSI/EEUM – Escola de Engenharia da Universidade do Minho,
Guimarães, Portugal
lpreis@dsi.uminho.pt

Abstract. The study of games in Artificial Intelligence has a long tradition. Game playing has been a fertile environment for the development of novel approaches to build intelligent programs. Multi-agent systems (MAS), in particular, are a very useful paradigm in this regard, not only because multi-player games can be addressed using this technology, but most importantly because social aspects of agenthood that have been studied for years by MAS researchers can be applied in the attractive and controlled scenarios that games convey. Diplomacy is a multi-player strategic zero-sum board game, including as main research challenges an enormous search tree, the difficulty of determining the real strength of a position, and the accommodation of negotiation among players. Negotiation abilities bring along other social aspects, such as the need to perform trust reasoning in order to win the game. The majority of existing artificial players (bots) for Diplomacy do not exploit the strategic opportunities enabled by negotiation, focusing instead on search and heuristic approaches. This paper describes the development of *DipBlue*, an artificial player that uses negotiation in order to gain advantage over its opponents, through the use of peace treaties, formation of alliances and suggestion of actions to allies. A simple trust assessment approach is used as a means to detect and react to potential betrayals by allied players. DipBlue was built to work with DipGame, a MAS testbed for Diplomacy, and has been tested with other players of the same platform and variations of itself. Experimental results show that the use of negotiation increases the performance of bots involved in alliances, when full trust is assumed. In the presence of betrayals, being able to perform trust reasoning is an effective approach to reduce their impact.

Keywords: Diplomacy · Strategy · Negotiation · Trust

© Springer-Verlag Berlin Heidelberg 2015
N.T. Nguyen et al. (Eds.): TCCI XX, LNCS 9420, pp. 179–200, 2015.
DOI: 10.1007/978-3-319-27543-7_9

1 Introduction

Since the beginning of Artificial Intelligence as a research field, game playing has been a fertile environment for the development of novel approaches to build intelligent machines. Besides puzzles and 2-player games (such as chess or checkers), for which contributions based on search and heuristics have been much successful, multi-player games make it harder to develop winning strategies. Multi-agent systems (MAS) are a useful paradigm for modeling such games, because of their natural fit into these scenarios. Furthermore, applying MAS research in games opens the possibility of applying social aspects of agenthood, which have been studied for years by MAS researchers, in the attractive and controlled scenarios that games convey.

Most approaches to game playing have been based mainly on (adversarial) search techniques and sophisticated domain-specific heuristics. Complex adversarial multi-player games pose new challenges to MAS research, given the fact that their search spaces are big enough to render ineffective any (current) approach based solely on search and heuristics. Of particular interest are those games in which a social dimension is included [14].

An example of the latter is Diplomacy, a military strategy multi-player simultaneous move board game, created by Allan B. Calhamer [1] in 1954. Its most interesting attributes include, according to Hall and Loeb [6], the enormous size of its search tree, the difficulty of determining the real strength of a position, and negotiation, whose support leverages the development of sophisticated players with an important competitive advantage. The fact that adversaries may negotiate throughout the game makes Diplomacy a very appealing sandbox for multi-agent research: while players are competing against each other, they must also cooperate to win the game. To do so, players may need to build trust, maintain relationships and negotiate deals, for which they may use argumentation techniques.

This work proposes an approach to the creation of a Diplomacy artificial player that takes advantage of negotiation and trust in order to enhance its performance. Our efforts focus on showing its competitive advantage, that is, on showing that by making use of social skills we are able to obtain an agent that is capable of surpassing its opponents. Our bot, *DipBlue*, works with the MAS testbed *DipGame* [4] and has been tested with another player of the same platform and with variations of itself (DipBlue archetypes).

The rest of the paper is structured as follows. Section 2 briefly describes the rules of Diplomacy and highlights the properties of the game that make it appealing for MAS research. Section 3 reviews related work on Diplomacy platforms and bots. In Sect. 4 we describe DipBlue's architecture and archetypes. Section 5 puts forward an experimental evaluation of DipBlue, presenting and discussing the obtained results. In Sect. 6 we draw some conclusions of the work done, and we point out some directions for future work.

2   Diplomacy: The Game

Diplomacy falls into the category of social games that explicitly allow collusion, a strategy deemed illegal in many real-world situations, such as markets and auctions. By including negotiation and allowing teamwork, Diplomacy offers the possibility of employing social skills when building intelligent playing agents, such as strategic negotiation, opponent modeling and trust reasoning.

Diplomacy action takes place in the beginning of the 20th century, in the years before World War I. Each player represents one of the following countries or world powers: England, France, Austria, Germany, Italy, Turkey and Russia. The main goal of the game is to conquer Europe, which is achieved by acquiring a minimum of 18 from a total of 34 *supply centers* throughout the map (see Fig. 1). During the game, each player commands its units in the map by giving them orders to *hold*, *move* to (i.e., *attack*) adjacent regions, or *support* other units actions (holds or moves of units from either the same or other players). Move actions to occupied regions originate conflicts (*standoffs*): the strongest unit (attacker or defender) wins a standoff, where strength is increased by backing up units with supports from other neighboring units. Some moves may invalidate other moves or cut supports. It is the conjunction of all orders that determines what actually happens in each round of the game[1].

Fig. 1. Standard Diplomacy map of Europe

Before each round of orders, players are able to communicate freely with each other in a negotiation phase. They can do so both publicly and privately, with the aim of establishing transient agreements. Although these conversations and arrangements are a huge part of the game-play (particularly in human tournaments), they hold absolutely no real power in the game itself: a player can commit to execute an action in exchange of information and decide not to fulfill

---

[1] Detailed rules of the game can be found in [1].

its part of the agreement once its individual goal has been achieved. Collective goals are thus always short-lived, given the most prominent individual goal of conquering the world.

Diplomacy is characterized by having no random factors (besides the initial assignment of world powers to players) and being a zero-sum game (as far as the final outcome of the game is concerned). However, the size of the game's search tree is enormous and impossible to search systematically even at low depths. In most other games, in order to address this problem the tree is pruned using heuristics that assess the state of the game at a given time and compare it with future game states. However, this technique cannot be directly applied to Diplomacy, given the fact that the game is played in a multi-agent partially observable environment [19], and thus not fully-deterministic from an agent's point of view – the chosen orders of a player are not necessarily effective, given its lack of knowledge about other agents' actions.

Solving problems by searching consists of representing the problem at hand as a search space (typically a graph) and employing a search technique to find a path through the graph. Heuristic search approaches (e.g. branch and bound, A*) try to take advantage of domain information to guide this process. Handling adversarial games consists of taking into account that certain moves in the path will be taken by the opponent, therefore building alternate decision layers in the search tree – the player does not have full control of the course of the game. Algorithms such as minimax and its variations are appropriate for handling these alternating moves kind of games, but still require the use of good heuristics to evaluate game states as accurately as possible.

Applying this kind of algorithms to Diplomacy is particularly challenging, not only because of its large state space, but also because it is hard to define appropriate game state evaluation functions. When attempting to create heuristics for Diplomacy (e.g. [6,20]), a player can be overlooked as a weak opponent when considering only the number and placement of its armies; and yet, when having strong alliances, a player can win the game or annihilate another player in a few turns. This makes the creation of an effective heuristic a difficult challenge.

Effective Diplomacy players need therefore to be developed using alternative (or complementary) approaches. The rich environment provided by Diplomacy promotes the development of bots capable of dominating their opponents through *negotiation*, which increases the need for *trust* reasoning capabilities to allow players to protect themselves.

3   Related Work

Sophisticated agent models have been developed over the years, including cognitive models for negotiation, argumentation, trust and reputation reasoning. In order to compare different advances in multi-agent systems, shared domains and rich environments are needed, providing challenging scenarios in which researchers can test their models. The Diplomacy game has been argued to provide such a testbed [3,5], adding also the possibility of putting together humans and agents in a complex interacting environment.

In fact, the Diplomacy game has been studied for a long time within the MAS research community. One of the first attempts to create a software agent for this game dates back to more than 25 years ago, by Kraus et al. [12]. Agent theories and architectures have been applied when developing Diplomacy agents. For example, Krzywinski et al. [14] have developed their Diplomacy agent following the well-known subsumption architecture.

Given the wide availability of human-based Diplomacy tournaments[2], some authors (such as Kemmerling et al. [10]) have devoted attention to develop human-like playing behavior, for which opponent modeling techniques play an important role. Developing agents that learn autonomously to play Diplomacy has also received attention from the research community. Shapiro et al. [20] have used temporal-difference learning and self-play to automatically develop a playing strategy, in an attempt to overcome the limitations of search-based approaches in the tremendous search space Diplomacy has. However, they focused in a non-cooperative version of the game, meaning that negotiation does not take place. On the other hand, Kemmerling et al. [11] applied an evolutionary algorithm-based planning approach for enhancing the performance of a Diplomacy agent that includes negotiation capabilities. However, given their bias on enhancing a believable bot, performance against the Albert benchmark (see below) is a bit weak. An optimization of a Diplomacy bot through genetic algorithms has also been done by de Jonge [9].

A short description of testbeds for Diplomacy follows, together with popular automated agents (bots) developed specifically for this game. We also review some of the main strategies used in the game, both in terms of evaluation heuristics and negotiation.

3.1   Diplomacy Testbeds

Although there are several different testbeds for MAS in general, there are a few specific for Diplomacy. The two most influential are briefly described here.

The Diplomacy Artificial Intelligence Development Environment (DAIDE[3]) assists the development of Diplomacy bots by taking care of all the logic concerning moves validation and the generation of resulting game states. It also provides a communication server that allows players to exchange messages between them during certain phases of the game. This communication server provides several layers of supported syntax in a way to allow for both simpler and more complex negotiating bots. The communication layers are referred to as *Press levels* and there are 15 distinct ones, ranging from the most basic (no communication at all) to the more complex level, in which free text negotiation is enabled. Both server and bots are written in C/C++.

DipGame[4] [5] is a testbed created at IIIA-CSIC that uses the DAIDE server to handle moves resolution and generation of new game states. Although DAIDE

---

[2] See e.g. http://www.playdiplomacy.com/.

[3] http://www.daide.org.uk/.

[4] http://www.dipgame.org/.

already supports communication, DipGame introduces its own server and creates a new communication syntax known as L Language. This language is composed of 8 levels, ranging from level 1 concerning deal negotiation, to level 7 foreseeing the use of argumentation. DipGame and its bots are implemented in Java. Additionally, DipGame provides an improved logging system and a web interface where anyone can play against some DipGame bots.

3.2 Diplomacy Bots

Some popular and relevant bots developed for Diplomacy are briefly mentioned here. These bots have different approaches, and some of them have been used as an inspiration during the creation of DipBlue.

Israeli Diplomat [12,13] was arguably the first attempt to create an automated Diplomacy player, by Kraus *et al.* It uses an architecture that distributes responsibilities according to the nature of the tasks. This architecture has served as an inspiration for other bots, such as the Bordeaux Diplomat. Israeli Diplomat has several well designed strategies to deal with heuristic search and negotiation.

The Bordeaux Diplomat [15] was created by Loeb and has a partitioned structure like the Israeli Diplomat, separating negotiation from solution search. The latter ignores the world power that owns each region and does an impartial evaluation of sets of actions by using a best first algorithm. The bot keeps a social relations matrix to determine the opponents that are more likely to betray.

DumbBot [16] is probably the most popular and common bot available for DAIDE. Even though it is not optimized and performs only a small tactical analysis, DumbBot performs relatively well, beating some attempts to create complicated heuristics and tactics. It does not perform negotiation of any sort – the only actions made are game-related orders. The bot has been the target of many studies and has been used as a benchmark for testing other bots. A replica of DumbBot was developed for DipGame [9], different only on the lack of support for a move called Convoy, which is not available in DipGame.

The Albert [21] bot was developed by van Hal and is, up until now, the best bot for DAIDE by far. It is the only Press Level 30 bot available. Because of its efficiency and high performance, it has been used as a benchmark by many researchers who try to outperform it.

BlabBot was created by Newbury [22] and builds on DumbBot by adding negotiation capabilities to it. BlabBot follows a "peace-to-all" strategy by sending peace offers to all players, decreasing the value of regions owned by players accepting those peace offers.

DarkBlade [18] is a no-press bot built by Ribeiro, which tries to combine the best tactics and strategies used by other Diplomacy agents. DarkBlade follows a modular architecture similar to Israeli Diplomat, and is modeled as an internal MAS, using so-called sub-agents.

HaAI [8] was developed by Johansson and Hååard, and uses a MAS structure inside the bot itself, in which each unit owned by the player is represented as an individual sub-agent. Each sub-agent tries to choose its own action according to what it considers to be the best option, while at the same time interacting as a team with the other sub-agents of the same player.

SillyNegoBot [17] is a DipGame bot developed by Polberg *et al.* and is an extension to the SillyBot, a bot similar to DumbBot. SillyNegoBot adds L Language level 1 communication and includes a BDI architecture. The bot has proven to be successful when matched with DumbBot but too naive when confronted with betrays. It uses the concept of personality with ratios for aggression/caution.

A few other works worth mentioning include an approach to optimize a Diplomacy bot using genetic algorithms [9], and a bot that takes advantage of a moves database, based on abstract state templates, providing the best set of actions for a given map and units with the goal of acquiring certain regions [2].

3.3   Strategies for Diplomacy

Evaluating board positions is crucial for effective Diplomacy playing. However, board evaluation is particularly complex in Diplomacy, both because of the partially observable environment a player is facing and the potential use of negotiation to establish temporary alliances between players. Had a player access to the private deals its opponents establish, a much more precise evaluation would be possible. Good references that describe strategies for Diplomacy include [6,22], besides all the information available online in sites such as DAIDE's. We limit ourselves to describe a few of the most often used ones.

The *province destination value* is used by DumbBot to assign a value to each region [9]. This metric takes into account the player that owns the region, and the amount of allied and enemy units in surrounding regions. The *blurred destination value* is a variation of the previous metric that spreads the value of a certain node to its neighbors. This way, the surrounding regions reflect that either the region itself is valuable or is near a valuable region. Values assigned to near regions can be obtained in a number of ways, e.g. by applying a Gaussian or linear blur.

Negotiation strategies often used in Diplomacy try to limit the search space by establishing cooperation agreements among players. However, when time comes such agreements may be simply ignored, and betrays come into play. This is why the establishment of an alliance does not *per se* comprise a real enhanced power to the players: the competitive advantage obtained by negotiating an agreement is based on the assumption of compliance, and thus any agreement is unstable in a zero-sum game like Diplomacy. Some of the main negotiation tactics that have been proposed in Diplomacy literature are briefly mentioned here. Many of these tactics are used by human players in real board games. However, they typically use concepts that are simple for humans but complicated for computers, such as small clues gathered just by looking at the opponents and the confidence the player has on other players.

The *peace-to-all* strategy is used in BlabBot, and tries to provide a certain level of security by quickly establishing alliances [22]. Players outside this set of alliances have a high chance of being eliminated, and the bot will progressively betray the player that is considered the most convenient to leave the allied group – usually the strongest one.

Back-stab is a tactic used by BlabBot for deciding when to betray alliances or for guessing when these will be betrayed by adversaries [22]. This tactic consists of keeping a threat matrix between the player and the opponents (and vice-versa): the higher the value, the more likely the player is to betray an alliance.

The *power-cluster* strategy is used to determine what world powers the player should ask for alliances and which ones to keep the longest. This strategy has evolved using clustering techniques over several games in order to identify which groups of powers have higher probability of succeeding, when allied.

4   DipBlue

In this section we present our own Diplomacy bot. DipBlue[5] is an artificial player built with the purpose of assessing and exploring the impact of negotiation in a game that natively relies on communication. Since the main difficulty when creating a Diplomacy bot is the size of the search tree, we have chosen to favor negotiation as DipBlue's main tool to gain advantage over its competitors, complemented with a simple form of trust reasoning to detect and react to betrayals.

4.1   Architecture

When designing DipBlue, we aimed at a flexible and easily extendible architecture. For that, a highly modular approach has been used, in which each module evaluates and determines, from its own perspective, the set of orders to apply in each turn. Figure 2 shows a class diagram comprising an overview of DipBlue's architecture, including two main components: Negotiator and Adviser (the latter is further explained in Sect. 4.3). Different advisers may be added as needed to the bot, exploiting its extensibility. This modular implementation also allows an easy customization of the bot, resulting in a vast array of possible configurations of bots that differ in their capabilities and behaviors. In Sect. 4.4 we discuss some of such configurations.

Figure 2 also shows the relation between one of the advisers and DumbBot, the bot it is based on. In other words, DumbBot could, in principle, be thought of DipBlue configured with a single adviser: MapTactician (see also Sect. 4.3).

The negotiation capability of DipBlue is materialized in the Negotiator component, responsible for handling received messages and for determining which messages are to be sent. Negotiation tactics are included in this component. The actual orders to be executed by each of the player's units, however, are dictated by Advisers. Any negotiated agreements that are to have an effect in further DipBlue actions need thus to be taken into account by some advisers (e.g. AgreementExecutor and WordKeeper in Fig. 2).

---

[5] DipBlue is named in honor of the chess-player supercomputer DeepBlue, and of the platform it is built to play on, DipGame.

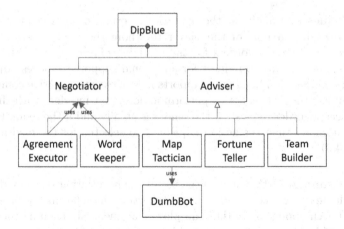

Fig. 2. DipBlue architecture

4.2 Negotiation and Trust

DipBlue is a negotiating bot with the ability to communicate in L Language level 1, whose format is explained in [5]. This layer of the language allows for three types of requests: *peace, alliance* and *order requests*.

Peace Requests. The most basic strategy of DipBlue consists of reducing friction with its opponents, in an attempt to make the game more predictable. For that, it makes use of peace requests, which reflect the intention for truce to occur among players and can be understood as a request for cease-fire or simply to achieve neutrality. In order to reduce the probability of conflict with as many players as possible, peace messages are sent to all negotiating players in the beginning of the game. DipBlue then opts to break truce with the player considered to be the least beneficial, taking into account the number of supply centers held by the other powers and the proximity between the power under analysis and DipBlue in the map.

Alliance Requests. In a more tactical level, alliance requests are handled by using two clusters of powers – allies and enemies – with the purpose of joining the efforts of the allied powers in order to defeat the enemies. DipBlue sends alliance requests to all players with whom it is in a state of peace, targeting the strongest non-ally power as an enemy. This results in a joint effort to eliminate the biggest threat at each phase of the game. Once the previously targeted enemy is weakened enough, the new strongest non-ally power is targeted, and so on. DipBlue accepts alliance requests from other players if they are in a state of peace and if the targeted enemy is not an ally itself.

Order Requests. Finally, at the operational level, an order request contains an order regarding a unit of the player to whom the request is sent. It has the purpose of suggesting orders for the receiving player's units. DipBlue uses these messages as a way to request for additional support to moves adjacent to allied units. Since the L Language supports messages with negative connotation, players can ask their allies not to perform actions that interfere with their own. DipBlue accepts order requests if the sender is an ally and if the requested order has a value higher than the action DipBlue had selected for the envisaged unit (see Sect. 4.3).

Trust Reasoning. Orthogonal to the use of the negotiation strategy described above is the maintenance of a *trust ratio* reflecting the relationship between the player and each opponent. Initially all players are neutral, meaning they have a trust ratio of 1. This ratio is converted into a friction ratio $Friction = 1/Trust$, used by the bot to decide on making alliances or to adjust the likelihood of fulfilling deals. It also determines when certain deals are accepted or rejected. The value of orders requested by other players is scaled with the trust ratio of the sender – players with a higher trust ratio have a higher probability of having their requests accepted.

Trust (or friction) ratios are updated during the course of the game. Events that decrease trust (and thus increase friction) include attacks and betrayals. Likewise, the lack of attacks by players in close distance or the fulfillment of agreements bring an increase on trust (and thus a decrease on friction). The magnitude of the impact of these events on trust depends on the current trust held by the player: trust in currently untrustworthy players is less affected; on the other hand, trustworthy players get a higher impact on their assigned trust value. This choice is meant to capture the competitive nature of the game of Diplomacy, emphasizing the role of betrayals during the game. An attack made by an ally (a currently trustworthy opponent) has a higher increase of friction than the same attack made by a current enemy. Given the nature of alliances in Diplomacy, which are not on solid ground and may suddenly be broken, with this approach we try to quickly capture such changes in the game.

Along with the trust ratio, a *state* that also reflects the current relationship is associated with each opponent. This state is originally *neutral* and may change to *war* or *peace* according to the trust ratio and the outcome of negotiations (namely peace and alliance requests). This state is used to enhance the impact of the trust ratio, by increasing its effect when assessing actions related with a given opponent. When a new alliance is started, all enemy player states are changed to war, thus reducing their trust ratio and increasing aggressiveness towards them.

4.3   Advisers

Advisers are the components of DipBlue that assess possible orders and determine what to do. Each adviser is independent, meaning that it can be used without the others, providing modularity and extensibility to the architecture. In the process of determining which actions to perform, the opinions of all employed advisers are taken into account.

A ranking of possible orders for each unit is created. In order to calculate the value assigned to each possible action, we use a weighted accumulation similar to a voting system, considering the numerical evaluation each adviser provides (see Eq. 1, where n is the number of advisers, w^i is the weight of adviser i and v^i_{Order} is the value Adviser i assigns to $Order$).

$$V_{Order} = \sum_{i=1}^{n} w^i . v^i_{Order} \tag{1}$$

While accumulating values, these can actually be either summed or multiplied, as for some advisers the assigned value has no meaning by itself (e.g. the probability of an order being successful), and should be interpreted as a scaling factor – the adviser is simply increasing or decreasing the importance of the order. This also means that the order of application of each adviser evaluation is important.

Finally, the best order for each unit is selected, ensuring they do not collide with each other. This verification is important because, for instance, if two units happen to attack the same region, a conflict arises and neither unit is successful, nullifying each other moves.

Initially, advisers have equal weights, which can then be adjusted in order to fine-tune the bot. Along with these weights, advisers themselves have intrinsic parameters that can be adjusted for obtaining different behavior variations. The adjustment of these parameters allows the creation of behavioral archetypes and personality, such as aggressive, naive, friendly or vengeful players. An optimization approach may be used to find out the optimal performance, following the approach in [9].

We now provide short descriptions of the advisers illustrated in Fig. 2.

MapTactician is the base adviser, serving as a starting point for all the following advisers to work upon. It is based on the behavior of DumbBot (see Sect. 3.2). This adviser performs an assessment of the map in terms of raw power, amount of enemy units and their positions, following a province destination value heuristic (see Sect. 3.3).

FortuneTeller takes into account the basic rules for resolving actions in Diplomacy to predict if an action will succeed, giving a probabilistic view of the evaluated move actions. Since Diplomacy has a complex set of rules with many exceptions and precedences between them, determining whether one action in a given set is going to be successful is not a trivial task. Given the branching factor of the search tree, it can also be quite time consuming. In order to alleviate this problem, FortuneTeller disregards the possibility of chain actions that may nullify each other, and thus often obtains optimistic probabilities of success.

The role of *TeamBuilder* is to promote support actions. Supports related with move actions that are highly ranked have their value increased, as a way to increase the probability of success of the move. Further in the process of choosing the actions for each unit, with this adviser a unit may abandon its highest ranked action to support instead some neighbor with a high need for support, particularly when the move of such neighbor has a value higher than the original action of the supporting unit. Increasing the weight of this adviser results in a higher cooperation in attacking moves, thus enhancing team play.

AgreementExecutor takes into account the deals made by DipBlue and decides how they should be performed. The value of each deal is assessed by taking into account the trust ratio of the deal's counterpart. Given the dynamics of the game, a deal may be proposed or accepted when the powers are in a friendly state but then be poorly rated because of a decrease of trust between both parties.

WordKeeper is the adviser in charge of reflecting the influence of trust/friction regarding each opponent. WordKeeper scales the value of the actions according to the trust ratio of the player the action is directed to. This way, the value associated with an attack to an ally is reduced, while the value associated with an attack to an enemy is increased.

4.4   Archetypes

Throughout the development of the DipBlue bot some distinct aspects were created, such as the ability to negotiate, propose deals and perform trust reasoning. In order to test some of these aspects individually, we have configured different DipBlue instances according to generic *archetypes*. Each archetype is defined by the set of advisers it uses and by the way the bot reacts to certain events, such as peace and action requests. Archetypes can be seen as different configurations of DipBlue, and were also defined to overcome the lack of DipGame bots available for testing purposes. With the exception of NoPress (see below), every other archetype described below uses all the advisers presented in Sect. 4.3.

Advisers are used as shown in Eq. 2. For the experiments reported in Sect. 5, all adviser weights have been set to 1. TeamBuilder, WordKeeper and FortuneTeller are used as scaling factors for the values obtained by MapTactition and AgreementExecutor.

$$V = ((((1 + MapTactician + AgreementExecutor)$$
$$\times TeamBuilder)$$
$$\times WordKeeper)$$
$$\times FortuneTeller) \tag{2}$$

NoPress is the most basic version of DipBlue. It does not perform negotiation nor trust reasoning of any kind. It is very similar to DumbBot in terms of capabilities. NoPress makes use of the MapTactician adviser.

Slave has the ability to communicate, although it does not take the initiative to start negotiations. Slave makes the same evaluation of actions as NoPress, automatically accepts every requests and follows them blindly (as long as they are executable). All agreements have higher priority as compared to the actions determined by the bot itself. This is the best bot to have as an ally.

Naive is endowed with the ability to propose deals of any supported kind to other players. When receiving incoming requests it has the ability to reason whether it should accept them based on a simple evaluation of both the request and the requesting player. Deals proposed by allies or players with very high trust ratio are likely to be accepted, while requests made by players the bot is in war with are almost always rejected. However, Naive lacks the ability to perceive when agreements are not fulfilled, and thus cannot be said to perform trust reasoning.

DipBlue is the more complete bot: it has the same setting as Naive with the addition of being able to perform trust reasoning. This allows DipBlue to detect hostile actions from other players and to assess how they fulfill agreements. Due to the use of trust ratios and both the AgreementExecutor and WordKeeper advisers, DipBlue is also capable of betraying other players. In Algorithm 1 a high-level specification of DipBlue's operation is listed.

Algorithm 1. DipBlue's high-level algorithm

Require: *gameState* {current state of the game}
    $\mathcal{A}$ {advisers to use}
    $\mathcal{X}$ {list of opponents}
    $\mathcal{P}$ {list of opponents in peace and their friction ratios}
    $\mathcal{W}$ {list of opponents in war and their friction ratios}
1: **for all** $op \in \mathcal{X}$ **do**
2:    *negotiatePeaceAgreement*($op, \mathcal{P}$)
3: **end for**
4: **while** *alive* **do**
5:    **switch** (*phase*(*gameState*))
6:    **case** *Spring, Fall*:
7:      *updatePeaceAgreements*($\mathcal{P}$)
8:      $hp \leftarrow highestPower(gameState)$
9:      *negotiateAlliance*($hp, \mathcal{P}, \mathcal{W}$)
10:     $\mathcal{O} \leftarrow selectMoveOrders(gameState, \mathcal{A})$
11:     *requestSupports*($\mathcal{O}, \mathcal{P}$)
12:    **case** *Summer, Autumn*:
13:     $\mathcal{O} \leftarrow selectRetreatOrders(gameState, \mathcal{A})$
14:    **case** *Winter*:
15:     $\mathcal{O} \leftarrow selectBuildOrRemoveOrders(gameState, \mathcal{A})$
16:    **end switch**
17:    *executeOrders*($gameState, \mathcal{O}$)
18:    **for all** $op \in \mathcal{X}$ **do**
19:      **for all** $o \in executedOrders(gameState, op)$ **do**
20:        **if** *isMoveTo*(o) **and** *target*(o) = *me* **then**
21:          *updateRatio*($op, \mathcal{P}, \mathcal{W}$)
22:        **end if**
23:      **end for**
24:    **end for**
25: **end while**

As mentioned in Sect. 4.2, the bot starts by proposing peace agreements to all adversaries (lines 1–3), and according to received responses updates the set $\mathcal{P}$ of opponents that are in peace.

When playing Diplomacy, in each season the players go through different phases, in the following sequence: spring, summer, fall, autumn and winter. Spring and fall are the so-called diplomatic phases, where players are able to negotiate cooperation (lines 6–11). DipBlue starts by revising peace agreements (line 7), taking into account what has happened in the previous phases. Friction ratios are updated and peace is broken for those opponents with a ratio above a given threshold. DipBlue will then select the highest power (line 8) as a target, proposing to all opponents currently in $\mathcal{P}$ an alliance to defeat it (line 9). Sets $\mathcal{P}$ and $\mathcal{W}$ are updated according to the responses received. Advisers in $\mathcal{A}$ are then used to evaluate and select move orders to be executed for each of the bot's units (line 10). Finally, for the selected orders support actions are requested from any opponent in $\mathcal{P}$ having a neighboring region (line 11).

Summer and autumn are phases where orders are executed (lines 12–13), and in case of standoffs, losing units need to retreat to an empty neighboring region or removed from the game. DipBlue uses its advisers in $\mathcal{A}$ to decide which retreat orders to execute for each dislodged unit (line 13).

Finally, winter is the phase where players earn additional units or lose exceeding ones according to the number of supply centers they occupy (lines 14–15). Again, DipBlue uses its advisers to decide where to place its newly acquired units or which units to remove (line 15).

After submitting its orders to the game for execution (line 17), DipBlue will analyze every executed order from its opponents (lines 18–24), and update ratios (line 21) for those players that have decided to attack it, i.e., that have executed move actions to one of its controlled supply centers (line 20).

It is important to emphasize that, for the sake of clarity, we have left outside this algorithm DipBlue's behavior in terms of responses to incoming peace, alliance or order requests. This behavior is informally described in Sect. 4.2.

5   Experimental Evaluation

When testing the performance of DipBlue, we were confronted with the lack of negotiating bots available for the DipGame platform. In order to overcome this problem, DipBlue archetypes have been tested in an incremental fashion. For that, a number of scenarios have been designed, as listed in Table 1.

In each scenario 70 games were made with the same specifications, and average and standard-deviation data has been computed. Following Diplomacy's rules, in each game 7 players are in play, which are randomly assigned to 7 different world powers.

Given the initially deterministic nature of DipBlue's archetypes, the usage of NoPress as the baseline version for measuring the performance of the remaining archetypes brought a significant number of games ending in a tie due to deadlocks. To avoid this, a simple randomization in the evaluation of moves was

Table 1. Testing scenarios.

#	Bots	Purpose
1	1x NoPress	Test the baseline version of DipBlue, which is theoretically
	6x DumbBot	equivalent to DumbBot
2	1x Slave	Test the performance of Slave when facing NoPress (no
	6x NoPress	negotiation)
3	1x Naive	Test the performance of Naive when facing NoPress (no
	6x NoPress	negotiation)
4	1x DipBlue	Test the performance of DipBlue when facing NoPress (no
	6x NoPress	negotiation)
5	1x Slave	Test the performance of the Naive archetype in the presence of a
	1x Naive	Slave, an agent that accepts and follows any proposed and
	5x NoPress	feasible deal
6	1x Slave	Test the performance of DipBlue in the presence of a Slave, an
	1x DipBlue	agent that accepts and follows any proposed and feasible deal
	5x NoPress	
7	1x Naive	Test the performance of DipBlue in the presence of a Naive, a
	1x DipBlue	deliberative team-player
	5x NoPress	
8	2x DipBlue	Test the performance of DipBlue when paired with an equal
	5x NoPress	player, which is also able to perform/detect betrayals

added to NoPress (consisting of adding between $-5\,\%$ and $5\,\%$ to the evaluation obtained by the MapTactician adviser). Tests showed that this randomization had no significant effect on performance in non-tied games, allowing us to nearly eliminate tie occurrences.

5.1 Crude Performance

As far as the rules of Diplomacy are concerned, the performance of a bot is determined by the position in which the bot ends the game. In games made with 7 equal bots, the average position is the 4th place.

By analyzing the average position obtained by NoPress in Scenario 1, which is 4.1, as shown in Fig. 3, it is possible to conclude that the performance of the bot is slightly lower than the performance of DumbBot. Because of this handicap and since NoPress is the foundation for all other bots and has no negotiation capabilities, the best way to measure the improvements of the remaining bots is to compare them with NoPress, rather than DumbBot.

In Scenarios 2–4 we test a Slave, a Naive and a DipBlue against 6 NoPress bots. In this setting, bots are not able to take advantage of negotiation; therefore, they rely on the same heuristics as NoPress, with the exception that DipBlue

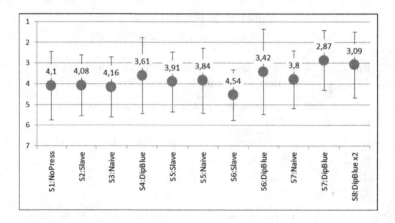

Fig. 3. Average and standard deviation of the final position of DipBlue archetypes in each scenario.

has trust reasoning based on the opponents actions. Both Slave and Naive were expected to perform exactly as NoPress and in fact they end up with a very similar average position. DipBlue, however, was expected to perform better than NoPress, given its ability to update friction ratios. It was actually capable of achieving a score of 3.61.

Scenarios 5 is used to assess how Naive and Slave work together, since Slave never rejects requests and Naive does not perform trust reasoning, this scenario should act as a reference point to assess the impact of a Slave when paired with other players, since a Slave may behave like a support player when allied to a player with the proper negotiation capabilities; Slave can be seen as a lever to other players and not as the subject of study itself. Although both bots have a slight performance increase, it is not significant to conclude that the communication between these bots was actually beneficial.

Scenarios 6 and 7 are used to evaluate how DipBlue interacts with communicating bots that do not perform trust reasoning. In Scenario 7, the Naive bot is able to accept a request from DipBlue and then decide not to follow the agreement. This will trigger DipBlue's trust reasoning capabilities and affect the way allies engage in future iterations. Results from this scenario show an obvious increase of performance for both bots, meaning that although they are able to betray each other, they are also able to successfully use the alliance in order to achieve mutual benefit, even more than when each of these bots is paired with a Slave (in Scenarios 5–6). From Scenarios 6–7 it also follows that DipBlue is able to get better results the more challenging its opponent archetype is. DipBlue's highest average position was obtained in Scenario 7: 2.87.

For Scenario 8, Fig. 3 shows the average position of both DipBlues. When paired with another instance of itself, DipBlue is able to detect betrayals and is vulnerable to be detected betraying. Therefore, when two instances of this bot are matched there is a high probability of conflict between two former allies.

The results highlighted by this scenario display an increase of performance when compared to Scenario 4 (with no negotiation taking place) and also when compared to Scenario 6 (where DipBlue was able to take advantage of a Slave bot). However, there is a decrease when compared to the performance of DipBlue in Scenario 7. While in that scenario DipBlue was able to betray alliances without repercussions, in Scenario 8 betrayals can be detected, which leads to a decrease of performance for the two bots.

From these observed results, we can state that negotiation capabilities allow DipBlue to enhance its results: all negotiation-able archetypes have shown a better performance than NoPress, except when used as a lever to other players, as shown with the Slave in Scenario 6. We can also conclude that mutual negotiation, where each player proposes agreements instead of just one player proposing and the other one following, works better than a "master/slave" scenario.

Furthermore, the implemented model for trust reasoning is an asset when in the presence of adversaries that are capable of negotiating and establishing agreements. Scenario 8 shows how both DipBlue's are able to achieve a good score while avoiding being exploited by the other one (something Slave and Naive are not able to do, in Scenarios 6 and 7, respectively).

Given the final goal of Diplomacy, which is to win the game, we also analyzed the winning capability of DipBlue. Figure 4 shows the percentage of wins of DipBlue in Scenarios 4, 6 and 7. Even though DipBlue has a better average performance with Naive in Scenario 7 (see Fig. 3), it is able to win more often by taking advantage of a Slave opponent. In fact, it is able to make slightly more wins when playing alone (as far as negotiation is concerned) in Scenario 4 than when paired with Naive, a much more challenging opponent than NoPress.

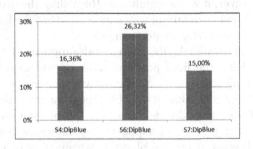

Fig. 4. Win percentage of DipBlue.

5.2   Correlation of Variables

In order to deepen the analysis of the obtained results, we have performed an inspection of dependencies between different variables, by defining correlation coefficients. All correlation coefficients regard the player position, which represents the ranking of the player. Therefore, negative coefficients mean the bigger the value of the variable the better the player's rank.

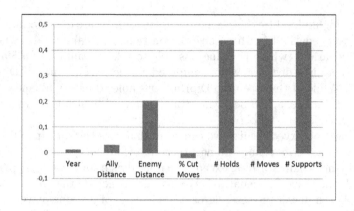

Fig. 5. Inverse correlation with final position of the bot.

Figure 5 displays the inverse correlation coefficients using aggregated data from all scenarios. Variables represent: number of years the game takes to end, distance to allies and enemies, percentage of moves cut (i.e. moves invalidated by other player moves), and the number of holds, moves and supports.

The correlation of the final position with the year in which games end is very reduced, meaning there is not a significant dependency between the length of the game and the performance of bots. The same applies to the percentage of moves that have been cut.

The correlation of the distance to allies has a low positive value, which indicates that a slight tendency of a gain in performance is obtained with the increase of the distance. However, it is not significant. Regarding the distance to enemies, Fig. 5 shows a significant correlation, which indicates that the farther the enemies are from the player, the better its performance.

Regarding the number of holds, moves and supports, these values display a high correlation with the final position. This is explained by the fact that by having more supply centers, a winning player also has a higher number of units, which in turn implies that it will be executing more actions. Therefore, the number of actions has a direct impact on the position of the player. The high correlation of the number of supports also indicates that teamwork is a good strategy, both when using the player's own units or recurring to those of its allies.

5.3   Impact of World Power

In Diplomacy, world powers determine to some extent the players' performance. One study made by Hunter [7] shows, for games played by humans, a discrepancy between powers both in no-press and in full-press games, as shown in Fig. 6. The results show an advantage of nearly double win percentage between France and Italy in both cases. Furthermore, one interesting result is that negotiation seems

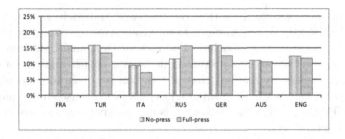

Fig. 6. Win percentages according to power in human games.

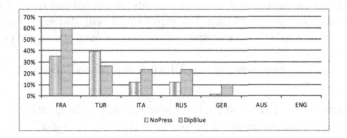

Fig. 7. Win percentages according to power in bot games.

to only benefit Russia – in every other case performance seems to be better in the no-press variant of the game.

While bot capabilities are far from being comparable to the performance of humans, it is an interesting exercise to compare power discrepancies when it comes to DipBlue's archetypes. Figure 7 shows such a comparison between NoPress (in 140 games ran with 7 NoPress bots) and the results obtained by DipBlue in Scenarios 4, 6 and 7. As shown, DipBlue seems to be able to enhance its performance in almost every case. As in human games, DipBlue achieves its best performance when playing France. Unlike human games, though, neither NoPress nor DipBlue are able to win any game when playing Austria or England. This is partially explained by the lack of convoys in DipGame, which makes it really hard for England to move through the Diplomacy board.

6   Conclusions and Future Work

Addressing multi-player games with cooperative strategies is a challenging domain for multi-agent systems. In this paper we have put forward an initial approach to develop negotiation-based agents for playing Diplomacy. The proposed modular architecture for DipBlue allowed us to test our bot using several different archetypes. The test scenarios had the purpose of highlighting certain aspects of the bots or their combination, producing results that allow to verify the validity of the proposed approach.

As a summary, we conclude that the proposed approach, DipBlue, successfully takes advantage of negotiation, as an alternative (or complement) to traditional solution search approaches. The lack of DipGame bots that are able to enter into negotiations has prevented us from a deeper analysis of our bots virtues. Nevertheless, we may say that negotiation is proven to be a very powerful approach in games where (temporary) cooperation between the players can take place. Furthermore, trust reasoning is a promising direction to address the breaking of agreements.

In the near future we would like to build, using DipBlue's architecture, different deliberative strategies for the game, better exploring negotiation features. This will also allow us to enrich our experiments by populating them with different negotiating bots for DipGame. Consequently, it will also enable us to make a deeper evaluation of the competitive advantages of each strategy as compared to the others.

Some promising improvements to DipBlue are planed, along the following lines.

Performance of World Powers. Bots performance varies greatly according to the world power they are assigned to. Reducing this effect would be beneficial to achieve a more stable and robust player, capable of having a good performance regardless of the world power assigned to it. However, as we have pointed out in our evaluations this is not an easy task, even for human experts.

Communication Capabilities. Negotiation strategies rely on communication. One of the most valuable improvements to be made is to increase the communication capabilities of the bot towards higher levels of the L Language.

Trust Reasoning. DipBlue performs a very simplistic trust reasoning. Being able to combine the previous actions of players with the current state of the game should enable a better assessment of the odds related with establishing or breaking agreements. Opponent modeling techniques and better game state evaluation functions should be used in order to better assess which are the expected strategic moves of opponents in each stage of the game.

Optimization. Following the approach described in [9], which applies genetic algorithms to optimize DumbBot [16], it should be possible to determine the best configuration of DipBlue, in order to achieve an optimal bot.

Learning. Using machine learning techniques, the bot can be endowed with the ability to learn from its previous experiences and opponents after a fair amount of games. This could be used to learn when to play each available action during the game or to improve negotiation tactics. Learning could also be used to predict the next opponent moves.

Finally, it is worth mentioning that the recent growth in MAS research applied to Diplomacy has given rise to a Computer Diplomacy Challenge in the Computer Olympiad 2015 event.

References

1. Calhamer, A.B.: The Rules of Diplomacy, 4th edn. Avalon Hill, Baltimore (2000)
2. Deyllot, R.J.G.: Diplomacy Base de Dados de Movimentos para Controlar Províncias. Master thesis, Universidade de Aveiro (2010)
3. Fabregues, A., Sierra, C.: Diplomacy game: the test bed. PerAda Magazine, towards persuasive adaptation, pp. 5–6 (2009)
4. Fabregues, A., Sierra, C.: A testbed for multiagent systems (technical report iiia-tr-2009-09). Technical report, IIIA-CSIC (2009)
5. Fabregues, A., Sierra, C.: Dipgame: a challenging negotiation testbed. Eng. Appl. Artif. Intell. **24**(7), 1137–1146 (2011)
6. Collier, N.: Repast: An extensible framework for agent simulation. Nat. Resour. Environ. Issues **8**, Article 4 (2001)
7. Hunter, E.: Solo percentages. http://www.diplom.org/Zine/W2003A/Hunter/Solo-Percentages.html. Accessed 29 July 2015
8. Johansson, S.J., Håård, F.: Tactical coordination in no-press diplomacy. In: Proceedings of the Fourth International Joint Conference on Autonomous Agents and Multiagent Systems - AAMAS 2005, p. 423 (2005)
9. Jonge, D.D.: Optimizing a diplomacy bot using genetic algorithms. Master thesis, Universitat Autònoma de Barcelona (2010)
10. Kemmerling, M., Ackermann, N., Beume, N., Preuss, M., Uellenbeck, S., Walz, W.: Is human-like and well playing contradictory for diplomacy bots? In: Proceedings of the 5th International Conference on Computational Intelligence and Games, CIG 2009, pp. 209–216. IEEE Press, Piscataway, NJ, USA (2009)
11. Kemmerling, M., Ackermann, N., Preuss, M.: Nested look-ahead evolutionary algorithm based planning for a believable diplomacy bot. In: Chio, C., et al. (eds.) EvoApplications 2011, Part I. LNCS, vol. 6624, pp. 83–92. Springer, Heidelberg (2011)
12. Kraus, S., Lehmann, D.: Diplomat, an agent in a multi agent environment: an overview. Technical report, Leibniz Center for Research in Computer Science (1987)
13. Kraus, S., Lehmann, D., Ephrati, E.: An automated diplomacy player. In: Levy, D., Beal, D. (eds.) Heuristic Programming in Artificial Intelligence: The 1st Computer Olympiad, pp. 136–153. Ellis Horwood Limited, Chinester (1989)
14. Krzywinski, A., Chen, W., Helgesen, A.: Agent architecture in social games - the implementation of subsumption architecture in diplomacy. In: Proceedings of the Fourth Artificial Intelligence and Interactive Digital Entertainment Conference, pp. 191–196 (2008)
15. Loeb, D.: Challenges in multi-player gaming by computers: a treatise on the diplomacy programming project. http://diplom.org/Zine/S1995M/Loeb/Project.html. Accessed 29 July 2015
16. Norman, D.: Dumbbot algorithm. http://www.daide.org.uk/index.php?title=DumbBot_Algorithm. Accessed 29 July 2015
17. Polberg, S., Paprzycki, M., Ganzha, M.: Developing intelligent bots for the diplomacy game. In: Proceedings of Federated Conference on Computer Science and Information Systems (FedCSIS 2011), pp. 589–596 (2011)
18. Ribeiro, J., Mariano, P., Seabra Lopes, L.: DarkBlade: a program that plays diplomacy. In: Lopes, L.S., Lau, N., Mariano, P., Rocha, L.M. (eds.) EPIA 2009. LNCS, vol. 5816, pp. 485–496. Springer, Heidelberg (2009)

19. Russell, S., Norvig, P.: Artificial Intelligence: A Modern Approach, 3rd edn. Prentice Hall, New Jersey (2009)
20. Shapiro, A., Fuchs, G., Levinson, R.: Learning a game strategy using pattern-weights and self-play. In: Schaeffer, J., Müller, M., Björnsson, Y. (eds.) CG 2002. LNCS, vol. 2883, pp. 42–60. Springer, Heidelberg (2003)
21. Hal, J.V.: Diplomacy ai - albert. https://sites.google.com/site/diplomacyai/home. Accessed 29 July 2015
22. Webb, A., Chin, J., Wilkins, T.: Automated negotiation in the game of diplomacy. Technical report, Imperial College London (2008)

Overcoming Limited Onboard Sensing in Swarm Robotics Through Local Communication

Tiago Rodrigues[1,2,3](✉), Miguel Duarte[1,2,3], Margarida Figueiró[1,2,3],
Vasco Costa[1,2,3], Sancho Moura Oliveira[1,2,3],
and Anders Lyhne Christensen[1,2,3]

[1] Bio-inspired Computation and Intelligent Machines Lab, 1649-026 Lisbon, Portugal
tlsrs@iscte-iul.pt
[2] Instituto de Telecomunicações, 1049-001 Lisbon, Portugal
[3] Instituto Universitário de Lisboa (ISCTE-IUL), 1649-026 Lisbon, Portugal

Abstract. In swarm robotics systems, the constituent robots are typically equipped with simple onboard sensors of limited quality and range.
In this paper, we propose to use local communication to enable sharing of sensory information between neighboring robots to overcome the
limitations of onboard sensors. Shared information is used to compute
readings for virtual, collective sensors that, to a control program, are
indistinguishable from a robot's onboard sensors. We evaluate two implementations of collective sensors: one that relies on sharing of immediate
sensory information within a local frame of reference, and another that
relies on sharing of accumulated sensory information within a global
frame of reference. We compare performance of swarms using collective
sensors with: (i) swarms in which robots only use their onboard sensors, and (ii) swarms in which the robots have idealized sensors. Our
experimental results show that collective sensors significantly improve
the swarm's performance by effectively extending the capabilities of the
individual robots.

Keywords: Multirobot systems · Evolutionary robotics · Situated
communication · Local collective sensing · Predator-prey task · Foraging

1 Introduction

Robots in large-scale decentralized multirobot systems, or *swarm robotics systems*, are typically equipped with simple and inexpensive sensors. This design
principle allows for the unit cost to be kept low, but limits the sensory capabilities of the individual robots (see, for instance, [11]). Many simulation-based
studies have disregarded limitations of real sensors [47], used simple communication to facilitate cooperation [22], or relied on indirect coordination through
stigmergy [3]. While simulation-based studies of robots with unrealistic sensors
can be used to analyze certain aspects of biological systems [17,46], the resulting controllers cannot be transferred to any real robotic system. Simple means
of communication, based on sound or color for instance, can be implemented

© Springer-Verlag Berlin Heidelberg 2015
N.T. Nguyen et al. (Eds.): TCCI XX, LNCS 9420, pp. 201–223, 2015.
DOI: 10.1007/978-3-319-27543-7_10

with off-the-shelf components in real robots [32], but they typically restrict the amount of information that robots can exchange in practice due to bandwidth limitations [7]. Still, bio-inspired approaches such as quorum sensing in bacteria [2,19], trophallaxis [40], and hormone-based communication [41] have been shown to be effective strategies to achieve coordination through relatively simple communication in multirobot systems.

Robots can alternatively be equipped with more complex, wireless communication hardware that enables direct transmission of binary data. In such scenarios, robots are able to transmit packets with relatively large amounts of information. Yet, in swarm robotics systems, wireless communication is typically only used to broadcast basic information, e.g. heading, location, or speed [10], in order to facilitate behaviors such as aggregation [24] and flocking [47].

In this paper, we show how sharing sensory information with neighboring robots can extend the sensory capabilities of the constituent robots in a swarm. The robots can sense shared information as readings from virtual sensors as though it had been sensed with onboard sensors. We call such virtual sensors *collective sensors*. We study two different approaches to the sharing of sensory information: (i) sharing of immediate sensory information on moving objects within a local frame of reference, and (ii) sharing of accumulated sensory information on static environmental features, such as a forageable resource, within a global frame of reference.

Sharing of sensory information within a local frame of reference can be achieved by using *situated communication* [27,43], where the signal that carries information also contains context, namely the relative direction and distance of the sender with respect to the receiver. Situated communication can be provided by widely available and relatively inexpensive equipment, such as the e-puck [33] equipped with the range & bearing board extension [27]. On the other hand, by using a global frame of reference, such as GPS coordinates, robots can share information about the location of static environmental features they have encountered.

We assess the performance of our approaches in two different tasks. In the first task, a swarm of robots must locate and consume preys in a classic predator-prey task, but where the robots share onboard sensory information using situated communication. A robot that receives information about a prey from a neighbor is able to estimate a prey's position within its own local frame of reference. The estimated positions of preys are then used to compute the readings for a receiving robot's collective sensors thereby effectively extending the robot's sensory range.

The second task is a foraging task, where robots must locate and transport resources to a predesignated area, the *nest*. Resources are clustered in the environment, and after a cluster has been depleted, a new one is placed at a random location. When a resource cluster is found by one of the robots, the robot stores the cluster's global location in memory. The robots then share information about the location of the resource cluster when they encounter other members of the swarm. In both tasks, we compare the performance of robots using collective sensors with robots that rely exclusively on onboard sensors, and with robots equipped with omnidirectional *ideal onboard sensors*, that is, unrealistic onboard sensors that have ranges equal to those of the collective sensors.

In this study, we use evolutionary robotics techniques [34] to obtain behavioral control for the robots. Evolutionary robotics techniques allow for the synthesis of control without the need for manual and detailed specification of the desired behavior [20]. Furthermore, an evolutionary approach to controller synthesis allows for self-organization of behavior, which is particularly relevant for the design of control systems for swarms of robots, where the individual behavior rules of each robot are difficult to derive [16].

While large number of studies on communication, coordination, and sensor fusion in multirobot systems have been conducted (see Sect. 2), the novel contributions of our study are: (i) the robots use collective sensors transparently as though they were onboard sensors, (ii) the reliance on local communication makes our approach scalable, and (iii) we use evolutionary robotics techniques which allows for the synthesis of control in which collective sensing becomes an integral part of the self-organized behavior.

Our goal is to maintain the desirable properties of natural swarm systems while simultaneously exploiting the unique capabilities of machines. We combine key properties of swarm intelligence systems such as scalability and robustness due to the exclusive reliance on decentralized control, with robots' capacity for low-latency and high-bandwidth communication in order to overcome limitations of the individual units' onboard sensory hardware.

2   Related Work

In this section, we review related work on communication and coordination in multirobot systems. We divide our review into three different sections: (i) traditional approaches to multirobot communication and coordination, (ii) bio-inspired approaches to communication and coordination in multirobot systems, and (iii) sensor fusion in multirobot systems.

2.1   Traditional Approaches to Communication and Coordination in Multirobot Systems

In one of the first studies on the relation between communication, coordination, and performance in multirobot systems, Balch and Arkin [1] evaluated the impact of global broadcast communication on performance. They studied teams of up to five robots performing three different tasks: two variants of a foraging task, and an exploration task. Several subsequent studies have proposed methods for effective task allocation and role assignment [5,23,36,42] in more complex scenarios. These methods are typically aimed at multirobot systems composed of relatively few, sophisticated robots. Prominent approaches include ALLIANCE [36] and market-based coordination [12]. ALLIANCE [36] is a software architecture for heterogeneous multirobot systems performing missions composed of loosely coupled tasks with possible ordering dependencies.

In ALLIANCE, coordination is achieved by giving each robot motivations (*impatience* and *acquiescence*) for performing subtasks that constitute a mission. Each robot performs task selection based on these motivations. While ALLIANCE is fully distributed, robots are assumed to be able to detect the actions of other team members by some means if broadcast communication is not available [36].

Market-based coordination, in which robots negotiate task allocation and role allocation through bidding, has been the subject of several studies [12,13,26]. Market-based approaches to coordination have been successfully demonstrated in a number of real robot scenarios including in an object manipulation task performed by a team composed of one watcher robot and two pusher robots [25], and in an exploration task performed by four PioneerII-DX robots [48].

Traditional approaches to multirobot coordination such as those discussed above, tend to be aimed at tightly coupled robot teams composed of relative few robots. Such approaches enable teams to coordinate their efforts effectively in dynamic environments, but require either global communication (even if unreliable and periodic, e.g. [42]), or means of detecting the actions of team members (e.g. [36]). For multirobot systems composed of a limited number of relatively sophisticated robots, these approaches have been successful. In robotic swarms, however, the individual units are typically assumed to be relatively simple, and control is decentralized [39]. Traditional approaches to coordination are therefore not applicable in the domain of swarm robotics.

2.2 Bio-inspired Approaches to Communication and Coordination

Biology has inspired several approaches to the communication and coordination of multirobot systems, ranging from bacteria [2,6,19] and insects [8,45], to flocking animals such as birds and fish [47].

Research on pheromone-laying ants has inspired a number of studies on multirobot coordination, in which robots, for instance, deposit a chemical substance [45] or use RFID tags [31] as pheromone. Payton et al. [37] used a transceiver to communicate the presence of virtual pheromone to nearby robots in a path-planning task. The authors assumed that a receiving robot was able to measure the intensity of the received signal in order to estimate the distance of senders. A similar approach was presented by Hoff et al. [28], but the communication requirements were simplified by not relying on signal intensities.

Local visual communication has been used to for a number of studies on bio-inspired communication and coordination, including the emergence of communication [21], detecting faulty members of a swarm [8], and coordination in swarm robotics systems [17]. Duarte et al. [17] studied the emergence of complex macroscopic behaviors observed in colonies of social insects, such as task allocation, communication, and synchronization. In a foraging scenario, robots were given the capacity to emit visual signals through changes in the robots' body color. The authors observed that explicit communication enabled complex behaviors to emerge, and the performance of the swarm was significantly higher than in scenarios in which robots could not emit signals.

Turgut et al. [47] studied self-organized flocking in a swarm of robots. The robots were equipped with a wireless communication module, which allowed the robots in the swarm to sense the headings of neighboring robots. By taking into account the robots' mean orientation, the swarm was able to achieve a robust flocking behavior. In a related contribution, Fredslund and Matarić [22] studied formation tasks in a swarm of robots with local sensing and minimal communication. The authors used robots equipped with a panning camera and infrared sensors. The robots' sensors allowed them to estimate the orientation and distance to other robots in the formation. No global localization was used, and only the chosen formation, robot IDs, and desired orientations were shared amongst robots.

In our study, we go beyond simple visual communication and transmission of basic parameters such as a robot's heading and speed. We process onboard sensory information, such as the estimation of the position of a target, and transmit it to neighboring robots. Collective sensors then use the received information to give the robot the capability to sense particular environmental features that would otherwise be beyond the range of the robot's onboard sensors.

2.3   Multirobot Sensor Fusion

Sensor fusion has been widely explored in team-based multirobot scenarios, such as robotic soccer, where tight coordination between robots plays a key role. Roth et al. [38] studied the construction of an individual world model, where robots could not communicate, and of a shared world model based on information exchanged between the robots. The authors' results showed that teams using a shared world model are more robust and successful at tracking dynamic objects than teams using individual world models.

Dietl et al. [14] developed a single-object tracking algorithm and a multi-object tracking algorithm based on Kalman filters [29]. The robots sent measurements from onboard laser range finders to an external computer that fused the received data, and then communicated the result back to the robots. The authors demonstrated that their technique led to robust and accurate behaviors. Karol and Williams [30] presented an approach based on an extended Kalman filter, which integrates observations of a single object from multiple robots and improves accuracy of the position and velocity estimates of objects. Their approach was successfully tested in a dynamic and uncertain robotic soccer task.

Pagello et al. [35] studied the problem of dynamic role assignment in a team of robots and the problem of sharing information to cooperatively solve an object tracking task. Two different types of communication were used: low-level communication based on stigmergy, and high-level direct communication for dynamic role assignment. The measurements of the object position and velocity were obtained from repeated observations or teammates' observations.

Although our approach shares some similarities with sensor-fusion techniques such as those presented in [35,44], there are significant differences: (i) collective sensors are indistinguishable from normal onboard sensors to the control program, (ii) information is only shared locally with robots within the communication range, therefore no global and/or centralized world model is used, which makes our approach scalable to large swarms of robots, and (iii) our behaviors are evolved, which allows for the synthesis of control in which collective sensing becomes an integral part of the self-organized behavior.

3 Methodology

In this study, we explore the potential benefits of sharing sensory information to extend the capabilities of the individual robots in swarm robotics systems. The proposed approach is based on the mutual sharing of information obtained from onboard sensors between neighboring robots. The shared information is then used to compute readings for collective sensors, which can give individual robots information that would not be available through their onboard sensors.

In our approach, robots can either share preprocessed information, such as the location of interesting features in the environment, or the raw readings obtained from simple sensors, such as, for instance, an infrared proximity sensor. The information is georeferenced and is broadcast to nearby robots, which can then use the information as though they had sensed it with their own onboard sensors.

It is important to georeference any shared sensory readings in order to maintain an accurate geospatial relationship between the receiving robot and the sensed environmental feature. If a local frame of reference is used, the relative location and orientation of the transmitting robot should be taken into account in order to calculate the readings for the collective sensors on the receiving robot. This can be achieved through communication means that implicitly embed such relative position information in the signals transmitted, namely situated communication [43]. If, on the other hand, a global frame of reference, such as GPS coordinates, is used, information about the location of an environmental feature is sufficient to compute readings for collective sensors on other robots.

The overhead of collective sensors is minimal, both in terms communication and computation. The information shared between the robots is either from simple sensors that produce a single or a few values, or high-level information such as the type and coordinates of an object. The information shared is thus limited to a few bytes per observation. Extracting high-level information from onboard sensor readings is necessary regardless of whether or not collective sensors are used. The overhead introduced by collective sensors for emitting robots is therefore limited to the broadcasting of high-level information. Robots receiving shared information, on the other hand, need only to translate the coordinates of the observations to their own frame of reference in order to compute the readings for their collective sensors.

A robot calculates the readings for its collective sensors taking into account its own location and orientation. Through fusion of information from several sources, collective sensors can provide robots with either longer range sensing, more accurate sensing, or both. In this study, robots exchange information regarding the relative position of perceived objects, effectively extending the sensory capabilities of the constituent robots in the swarm. A key difference with respect to previous studies is that, in our approach, locally shared information is used to compute readings for the collective sensors, which, in turn, are used by the robots as if they were onboard sensors.

We compare two variants of sensory sharing: (i) sharing of *immediate* sensory information, and (ii) sharing of *accumulated* sensory information. Sharing of immediate sensory information is used when information concerning moving objects may quickly become obsolete, when dead-reckoning is subject to cumulative errors, and when global localization is not available. If global localization is available, however, sensory information accumulated over longer periods of time can be shared, provided that the environmental features of interest remain static.

For the experiments presented in this paper, each robot is controlled by an evolved artificial continuous-time recurrent neural network [4] with a reactive layer of input neurons, one hidden layer with ten neurons, and one layer of output neurons. The input layer is fully connected to the hidden layer, which, in turn, is fully connected to the output layer. The input layer has one neuron for each sensor, and the output layer has one neuron for each actuator. The neurons in the hidden layer are fully connected and governed by the following equation:

$$\tau_i \frac{dH_i}{dt} = -H_i + \sum_{j=1}^{S} \omega_{ji} I_i + \sum_{k=1}^{H} \omega_{ki} Z(H_k + \beta_k) \tag{1}$$

where τ_i is the decay constant, H_i is the neuron's state, S is the number of sensors, ω_{ji} the strength of the synaptic connection from neuron j to neuron i, H is the number of hidden neurons, β the bias terms, and $Z(x) = (1+e^{-x})^{-1}$ is the sigmoid function. β, τ, and ω_{ji} are genetically controlled network parameters. The possible ranges of these parameters are: $\beta \in [-10, 10]$, $\tau \in [0.1, 32]$ and $\omega_{ji} \in [-10, 10]$. Circuits are integrated using the forward Euler method with an integration step-size of 0.2 and neural activations are set to 0 when the network is initialized.

Each generation in the evolutionary algorithm is composed of 100 genomes. After all the genomes are evaluated, an elitist approach is used: the top five genomes are selected to populate the next generation. Each of the top five genomes is used as the parent of 19 offspring. An offspring is created by applying a Gaussian noise to each gene with a probability of 10 %. The 95 mutated offspring and the original five genomes constitute the next generation. For our experiments, we use JBotEvolver [18], an open source, multirobot simulation platform and neuroevolution framework.

4 Sharing of Immediate Sensory Information

In this section, we present results of experiments involving robots equipped with collective sensors based on immediate sharing of onboard sensory information. In a predator-prey task, predator robots share information about the location of preys through local, situated communication.

4.1   Experimental Setup

We evaluate the immediate sharing approach in a predator-prey task where a group of robots (the predators) with limited onboard sensing must locate and consume a number of moving preys. The environment is square-shaped, with a size of $10\,m\times10\,m$, surrounded by walls. The robots start each experiment in the center of the environment, while the preys are placed in random locations sampled from a uniform distribution. A robot can capture and consume a prey by touching it. Whenever a prey is consumed, a new prey is placed at a random location in the environment, and the total number of preys is thus kept constant throughout an experiment.

We use small (10 cm diameter) differential-drive robots, loosely modeled after the e-puck [33]. The maximum speed of the robots is $10\,cm/s$. The set of sensors is composed of: (i) two onboard prey sensors with a range of $0.8\,m$, (ii) four collective prey sensors with a range of $3\,m$, (iii) four robot sensors with a range of $3\,m$, and (iv) four wall sensors with a range of $0.5\,m$. All the sensors have an opening angle of $90°$. The collective prey sensors, the robot sensors, and the wall sensors are oriented toward the angles $0°$, $90°$, $180°$ and $270°$ with respect to the front of the robot, while the two onboard prey sensors are oriented toward angles of $15°$ and $-15°$ (see Fig. 1). Consequently, the onboard prey sensors overlap by $60°$ and jointly cover a section of $120°$. The fact that the two onboard prey sensors overlap was found to help the robot to locate and pursue preys in a set of preliminary experiments. The onboard prey sensors could be implemented on real robots, based on inputs from a front-facing onboard camera, for instance, by dividing the captured images into two overlapping regions.

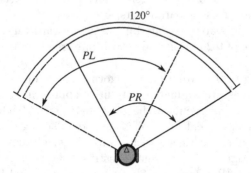

Fig. 1. Location and field of view of the two onboard prey sensors, where PL indicates the area sensed by the left onboard prey sensor and PR the right onboard prey sensor. The onboard prey sensors have a range of $0.8\,m$ and opening angle of $90°$. Together they cover a $120°$ wide section, and overlap by $60°$.

Readings for the collective sensors are computed based on estimates received from nearby robots that are detecting a prey with their onboard sensors. The relative position of the prey is calculated taking into account the relative distance

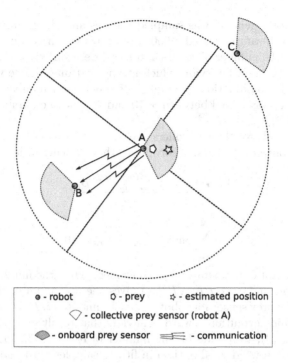

○ - robot   ○ - prey   ☆ - estimated position

▽ - collective prey sensor (robot A)

◖ - onboard prey sensor   ⇒⇐ - communication

Fig. 2. An illustration of how collective sensors work. When robot A senses a prey with its onboard sensors, it processes the sensed information generating a estimate of the prey's position and broadcasts the estimate to nearby robots, in this case robot B. Since robot C is outside robot A's communication range, robot A's estimates do not reach robot C.

and orientations of the nearby robots, as well as the prey's relative location with respect to the robot that is detecting the prey. If estimates are received from multiple robots, it becomes possible to triangulate the position of the prey. Otherwise, an a priori estimate of 50 cm is used for the distance between the prey and the robot that is detecting it. The a priori estimate of 50 cm corresponds to 10 times the radius of the robot. The sharing of information is limited to the range of the local, situated communication technology. In this study, the range of both the collective sensors and of the local, situated communication is 3 m. An illustration of the collective sensors can be seen in Fig. 2.

The preys are able to move at a speed of 15 cm/s, which is 1.5 times faster than the robots. Each prey has two wall sensors and four robot sensors. The prey remains still whenever no nearby robot is detected. If a robot is detected and if no robot is directly in front of the prey, the prey moves forward at full speed. If a prey detects a robot in front of it, the prey randomly turns either left or right until it is able to move forward. After a prey escapes, it remains static until a nearby robot is detected again.

The fitness of a genome was sampled 9 times and the mean fitness is used for selection. Each sample lasted 5,000 time steps, which is equivalent to 500 s. In each sample, the number of robots and preys were varied in order to promote the evolution of general behavior, which means that one sample was conducted for each possible combination of number of robots and number of preys. The number of robots was varied between 5, 10 and 20, and the number of preys was varied between 2, 5 and 10.

In order to evaluate the controllers, we rewarded robots for moving close to preys and for consuming preys, according to the following fitness function:

$$F = \frac{NP + \sum_{t=0}^{\text{time steps}} B_t}{NR} \tag{2}$$

$$B_t = \sum_{r=0}^{NR} \max(PL_{r,t}, PR_{r,t}) \cdot 10^{-5} \tag{3}$$

where NP is the total number of preys consumed, NR is the number of robots in the environment, and $\max(PL_{r,t}, PR_{r,t})$ gives the maximum of the readings of the left and right prey sensor for robot r at each time step t. B_t is a bootstrapping term used to guide evolution toward behaviors that result in robots being close to preys. In Eq. (3), we divide by the number of robots, NR, in order prevent biasing evolution toward local optima in fitness samples with many robots.

We ran experiments for three different setups: (i) a setup in which robots have collective sensors, that represents our approach, (ii) a setup in which robots use only their onboard sensors, and (iii) a setup in which robots have ideal sensors that let them sense preys up to a range of 3 m, which is equal to the range of the collective sensors. We ran 20 evolutionary runs in every setup, each lasting 500 generations. We conducted a post-evaluation of the genome that had obtained the highest fitness in each run with a total of 900 samples, 100 for each combination of numbers of robots and preys.

4.2   Results and Discussion

Performance. Figure 3 shows the average fitness scores of the highest scoring controllers evolved in the collective sensors setup, onboard sensors setup, and ideal sensors setup, respectively. Each boxplot summarizes the results obtained in 900 post-evaluation samples, from 9 different configurations of number of robots and preys. The results show that the highest-performing controllers evolved in the collective sensors setup outperformed the controllers in the onboard sensors setup, and underperformed the controllers in the ideal sensors setup. The average fitness obtained by the highest-performing controllers in post-evaluation for the collective sensors setup, onboard sensors setup, and ideal sensors setup corresponds to 0.43 ± 0.24, 0.12 ± 0.04 and 0.72 ± 0.07, respectively, and in terms of preys consumed to 5.36, 1.61 and 8.81, respectively (see Table 1).

When comparing the performance of the controllers evolved in the collective sensors setup with those evolved in the ideal sensors setup, the latter achieved a

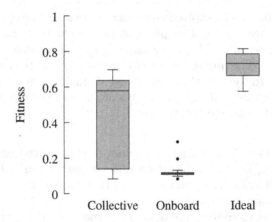

Fig. 3. Boxplot of the post-evaluated fitness scores achieved by the highest-scoring controllers in 20 evolutionary runs conducted in each of the setups. Each box summarizes results from 900 post-evaluation samples, and comprises observations ranging from the first to the third quartile. The median is indicated by a bar, dividing the box into the upper and lower part. The whiskers extend to the farthest data points that are within 1.5 times the interquartile range, and the dots represent outliers.

Table 1. Mean fitness obtained and number of preys consumed by the highest-performing controllers in post-evaluation of the controllers evolved in the collective sensors setup, onboard sensors setup, and ideal sensor setups.

	Fitness	Preys consumed
Collective	0.43 ± 0.24	5.36 ± 2.92
Onboard	0.12 ± 0.04	1.61 ± 0.51
Ideal	0.72 ± 0.07	8.81 ± 0.84

higher fitness, which can be explained by the fact that the collective sensors need at least one robot detecting a prey with its front prey sensors to be able to share the prey's relative position with nearby robots. In the ideal sensors setup, the prey sensors have a range of 3 m instead of 0.8 m as in the other two setups, and detect preys in any direction with respect to the orientation of the robot. The differences between robots with collective sensors and robots with ideal sensors translate into a mean difference of preys consumed in post-evaluation of 3.45 preys.

In order to evaluate the robustness, adaptivity, and scalability of the solutions evolved, we evaluated the controllers from the highest-performing evolutionary run using collective sensors in an environment where the principal factors — area of the arena, number of preys and robots, were scaled by a factor of five, resulting in an arena of 22.3 m×22.3 m (500 m^2), 50 preys and 100 robots. The time given to the robots for performing the task was kept the same as in the original setup (500 s). The evolved controllers were able to disperse well, locate and consume an average of 44.6 preys in post-evaluation experiments, three times the average number of preys consumed by the controller in the original setup (14.8 preys). The number of preys consumed was only three times higher and not five, due to the fact that the average distance from a robot to the wall is longer in the enlarged arena, and robots often need to trap preys in corners or along walls before they can catch them.

Behavior. We identified two sub-behaviors in the evolved behaviors for the immediate sharing approach, based on observation: a search behavior and a trap/consume behavior. The preys are faster than the robots, which means that the robots often have to trap a prey before they can catch it. A prey can become trapped if it moves close to a wall or into a corner, and two or more robots are following it closely. Alternatively, three or more robots can trap a prey without the aid of walls by approaching from different directions.

In the collective sensors setup, 15 out of 20 runs evolved the same type of behavior: at the start of a trial, the robots disperse in an outward spiraling motion in order to find preys. The robots then try to chase preys toward the corners or a wall, either in groups or alone. When a prey is consumed or escapes, the robots disperse again to cover a larger area. An example of this behavior can be seen in Fig. 4.

The most significant difference found in the behaviors are in the extremes of number of robots, that is, between samples where 20 robots are present and samples where only five robots are present. When the density of robots is high, the robots tend to disperse evenly and when a prey is seen, they quickly aggregate around the prey with nearby robots. On the other hand, when the density of robots is low, aggregation near a prey is slower since the robots tend to be more distant from each other, forcing each robot to try to trap a prey alone, or try to maintain the prey in view until another robot gets within communication range.

In five evolutionary runs, the highest-performing controllers of the collective sensors setup display a behavior in which the robots move backwards. The behavior in which robots move backwards corresponds to a poor local optimum in which evolution became stuck in early generations. In this optimum, the robots tend to have a relatively fixed motion pattern that, by chance, can cause one or more preys to be trapped in corners and then consumed.

In the highest-performing behaviors evolved in the onboard sensors setup, the robots start by dispersing in an outward spiraling motion to find preys. When a prey is found, the robots attempt to pursue it until another robot is able to detect the same prey with its onboard sensors. The highest-performing controllers of the ideal sensors setup have a different behavior: since the robots are almost always able to see a prey, they simply pursue and try to consume the closest prey, and hence spend little or no time searching for preys.

Controllers evolved in the collective sensors setup tend to have a performance closer to the performance observed in controllers evolved in the ideal sensors when the number of preys is higher than the number of robots (Fig. 5).

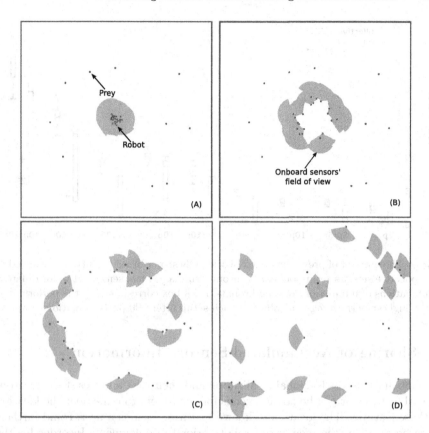

Fig. 4. Example of a high-performing controller evolved in the collective sensors setup, in a sample with 20 robots and 10 preys. The robots start at the center of the arena (A), and then disperse in an outward spiraling motion in order to find preys (B). When preys are detected (C), the robots chase the preys toward the corners or walls, either in groups or alone (D).

Since the robots in the ideal sensors setup can sense preys up to a distance of 3 m, they often tend to follow different preys, which sometimes makes it difficult to trap and consume them. The robots in the collective sensors setup, on the other hand, tend to follow fewer preys in larger numbers, due to their reliance on collective sensors.

In summary, our results show that sharing immediate sensory information offers significant performance benefits over robots that only use limited onboard sensing in the predator-prey task. In the approach demonstrated in this section, robots only share immediate sensory information, that is, the location of a particular environmental feature (e.g. a prey) if they are currently observing it. However, as we detail in the next section, sensory information accumulated over time can be shared in scenarios where global localization is available and where environmental features of interest remain static.

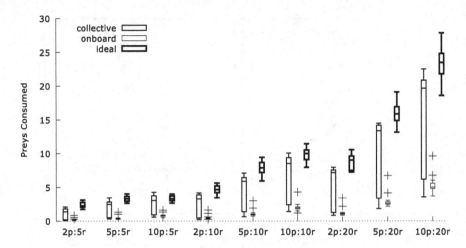

Fig. 5. The number of preys consumed by the highest-performing controllers evolved in the collective sensors setup, onboard sensors setup, and ideal sensor setup for different combinations of number of preys and robots. Each box corresponds to the performance of the highest-performing controllers in each setup after 100 post-evaluation samples.

5   Sharing of Accumulated Sensory Information

If robots have access to global localization and their task involves static environmental features, it can be advantageous for the robots to remember the location of those features. The location of known features, such as a nest, could furthermore be saved in the memory of robots prior to deployment. In order for the robots to be able to use stored locations of environmental features, we endow the robots with virtual sensors, which can compute the relative heading and distance of the robot to the features stored in memory. Since readings are computed based on the robot's current location and the content of its memory, such virtual sensors are not physically limited in terms of range as typical onboard sensors are.

By communicating with one another, the robots can share stored locations of features, allowing robots that receive a feature's location to sense it with their virtual sensors. We make the distinction between collective sensors and virtual sensors, depending on whether or not the locations stored in memory are shared with other robots. Virtual sensors thus become *collective* when robots share the content of their memory.

In this section, we experiment with a foraging scenario in which robots must search for resources to then transport back to a nest. We assume that resources are static, and that robots know their own current position and orientation within a global frame of reference. When a robot finds a cluster of resources, it stores the cluster's location in memory and then shares it with other members of the swarm as they get within communication range.

5.1   Experimental Setup

The foraging environment is unbounded and a cluster of resources is initially placed at a random location, sampled from a uniform distribution, within a square area measuring 120 m by 120 m around the nest. Resource clusters contain between 5 and 15 units when placed in the environment. A robot can collect and transport one unit of resource at a time. When the number of units in a cluster reaches 0, it is removed, and a new cluster with a random number of units in the interval $[5, 15]$ is placed in the environment at a random location.

A total of 10 robots start in the nest. We use differential-drive robots, with a diameter of 1 m and with a maximum speed of 1 m/s. The robots can communicate with each other at a range of up to 20 m. Each robot has a set of sensors composed of: (i) two onboard resource sensors with a range of 5 m, (ii) four virtual resource sensors with a range of 120 m, (iii) four robot sensors with a range of 20 m, (iv) four nest sensors with a range of 120 m, and (v) one sensor indicating if a robot is currently carrying a unit of resource or not. All the sensors have an opening angle of $90°$. The virtual resource sensors, the robot sensors, and the nest sensors are all oriented toward the angles $0°$, $90°$, $180°$ and $270°$, while the two onboard resource sensors are oriented toward angles of $15°$ and $-15°$. The onboard resource sensors allow robots to sense the location of the closest resource cluster and the number of units left in the cluster. The onboard resource sensors overlap by $60°$ and cover a section of $120°$. In terms of actuators, each robot is equipped with two wheels and one gripper that enables the robot to pick up and carry a unit of a resource cluster.

Whenever a robot detects a resource cluster with its onboard sensors, it records the current time, the number of units in the cluster, and the cluster's global position. Robots thus timestamp observations of resource clusters, and include the time of observation when information on the location of known clusters is communicated to other robots. Since resource clusters eventually are depleted and disappear, we make the period of time that information is kept in memory proportional to the number of units of resource in the cluster. The period of time is calculated by multiplying the observed number of units of resource by 20 s.[1] Robots update their registered observations by comparing the timestamp of the received location with the timestamp of the location stored in memory. Information stored in memory is updated if the received information is newer than the stored one. Readings for the virtual sensors are computed based on information stored in memory, namely the location of the closest resource cluster and the last known number of units remaining in that cluster.

We ran experiments for three different setups: (i) a setup in which robots have collective sensors, which represents our approach, (ii) a setup in which robots use only their onboard sensors, but store the global position of detected clusters and use virtual sensors, and (iii) a setup in which robots have ideal sensors that let them sense resources up to a range of 120 m, which is equal to the range of

---

[1] In preliminary experiments, we found that performance results are robust to moderate increases in the time-multiplication factor (up to 50 s).

the collective sensors. We ran 30 evolutionary runs in every setup, each lasting 500 generations.

The fitness of a genome was sampled 30 times and the mean fitness was used for selection. Each sample lasted $5,000$ time steps, which is equivalent to $500\,s$. In order to evaluate the controllers, we rewarded robots for: (i) moving close to a resource cluster when not carrying a resource, (ii) moving towards the nest when carrying a unit of resource, and (iii) deploying resources in the nest, according to the following equations:

$$F = FR + \sum_{t=0}^{\text{time steps}} B_t \qquad (4)$$

$$B_t = \sum_{r=0}^{NR} \begin{cases} \left(\frac{S - DC_{r,t}}{S} \cdot 10^{-6} \right) & \text{if not carrying a resource} \\ \\ \left(\frac{S - DN_{r,t}}{S} \cdot 10^{-5} \right) & \text{otherwise} \end{cases} \qquad (5)$$

where FR is the total number of units of resource foraged, NR is the number of robots in the environment, S is the side length of the foraging area (120 m), $DC_{r,t}$ is the distance of robot r to the closest resource cluster at time step t, and $DN_{r,t}$ is the distance of robot r to the nest at time step t. B_t is a bootstrapping term used to guide evolution toward behaviors that result in robots being close to resource clusters when not carrying a unit of resource, and close to the nest when carrying resources.

5.2   Results and Discussions

Performance. After all evolutionary runs had finished, we conducted a post-evaluation with a total of 100 samples of the genome that had obtained the highest fitness in each run. The performance of the highest-scoring controllers of the collective sensors setup, onboard sensors setup, and ideal sensors setup for the accumulated sharing approach are shown in Table 2, which contains the results of the 100 post-evaluation samples, and lists the average fitness scores, which reflect an approximation of the units of resources collected, obtained in each setup. The average fitness acquired in post-evaluation by the highest-scoring controllers of the collective sensors setup was 9.65 ± 1.23, outperforming the controllers with only onboard sensors, with an average fitness of 6.47 ± 0.64, while underperforming the ideal sensors controllers, with an average of 37.84 ± 1.65.

The lower performance of the controllers evolved in the collective sensors setup, when compared to the ones evolved in the ideal sensors setup, can be explained by the fact that it is necessary that at least one robot has encountered a resource cluster in order for its location to be shared with neighboring robots. Since the robots' onboard sensors have a relatively short range (5 m), robots had to spend a significant amount of time searching for resource clusters.

In the ideal sensors setup, robots are able to detect resources in any direction within a 120 m range, meaning that robots almost never have to search for a resource cluster, and can instead go to one directly, hence the outstanding performance.

Table 2. Mean fitness and standard deviation obtained by the highest-performing controllers in post-evaluation of the controllers evolved in the collective sensors setup, onboard sensors setup, and ideal sensor setup.

	Fitness	St. dev
Collective sensors setup	9.65	±1.23
Onboard sensors setup	6.47	±0.64
Ideal sensors setup	37.84	±1.65

Behavior. Both the collective sensors setup and onboard sensors setup evolved similar behaviors to locate resource clusters. We observed that robots start by dispersing in an outward spiraling motion around the nest, in what can be described as a search behavior. This behavior is repeated each time robots cannot detect any resource cluster with their onboard sensors or with their virtual sensors.

The behavior for the actual foraging differs between setups. In the collective sensors setup, when a resource cluster is detected, the location of the cluster is shared with the robots within communication range, facilitating aggregation and foraging of the cluster. When a robot is transporting a unit of resource to the nest, the cluster's location is transmitted to other robots that move within its communication range, enabling robots to detect the resource cluster with their collective sensors. Robots update the content of their memory when a newer observation is received. Since robots store resource clusters' locations in memory, they can skip the search phase and move directly to the stored location of a cluster after depositing a resource in the nest. An example of the behavior evolved in the collective sensors setup can be seen in Fig. 6.

Robots from the onboard sensor setup, which are not equipped with collective sensors, perform circular movements around the nest until a resource is detected with their onboard sensors. In 16 out of 30 runs, the robots evolved a behavior in which they forage individually without any evident collaboration. However, in 11 out of 30 runs, robots evolved a behavior in which, after locating a resource cluster, a robot remains near the cluster, which attracts nearby robots. This collaborative strategy, proved to be less efficient than the one found on the collective sensors setup, since robots are spread around the environment trying to locate resources on their own. In the three remaining runs, one evolved a behavior in which robots use the individual foraging strategy when a resource cluster is located closer to the nest and the collaborative strategy when resources are distant from the nest. In the two remaining runs, we observed both the collaborative behavior and the

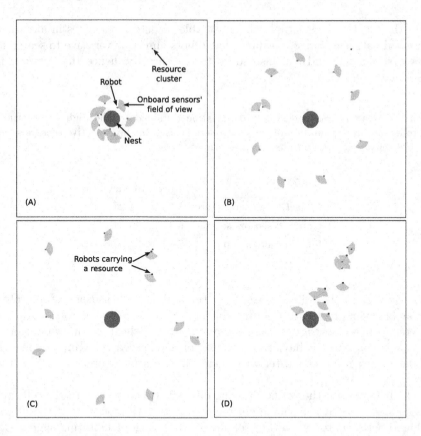

Fig. 6. Example of a high-performing controller evolved in the collective sensors setup. The robots start at the center of the unbounded environment (A), and then disperse in an outward spiraling motion in order to find a resource cluster (B). When a robot detects a cluster, it collects a resource and transports it to the nest (C), then other robots start to move toward the cluster (D). Robots carrying a resource unit are shown in black for clarity.

individual behavior, but the choice of behavior did not appear to depend on the distance between a resource cluster and the nest.

The performance difference between the collective sensors setup and the ideal sensors setup is an indication that the search phase of the task is the most unproductive and time-consuming subtask. The robots equipped with ideal sensors can effectively skip the search phase, since they can immediately sense the location of a resource cluster. This results suggest that there may be significant benefits in combing our approach with heterogeneity in a swarm's onboard sensors. Some robots could, for instance, be equipped with better, long-range sensors than the rest of the swarm. Robots with long-range sensors could then more efficiently find resource clusters and share their location with the robots with lower-range sensors.

6   Conclusions

In this paper, we explored a novel approach in which robots share information and readings from their sensors with neighboring robots to overcome the often limited capabilities of onboard sensors in robotic swarms. Robots are able to georeference environmental features sensed with their onboard sensors and then share information locally with robots within communication range. Georeferencing can be achieved using either a local frame of reference and shared using situated communication, or a global frame of reference where the coordinates of a particular environmental feature are shared. Shared information is then used by receiving robots to compute readings for their collective sensors.

We evaluated our approach in two different scenarios: (i) in a predator-prey scenario, in which the estimated positions of detected preys were immediately communicated to neighboring robots, and (ii) in a foraging scenario in which the locations of resource clusters were stored in robots' memory, and therefore could be shared post observation. Received information was used to compute the readings for collective sensors, thereby effectively allowing robots to sense environmental features at greater distances.

Our experimental results showed that in the two scenarios, swarms using collective sensors achieved a higher performance than swarms in which the robots relied exclusively on their onboard sensors. In certain setups, the performance of swarms using collective sensors with immediate sharing even approached the performance of swarms in which robots were equipped with ideal sensors. In the context of the CORATAM project [9], we have conducted successful preliminary experiments with collective sensors in a swarm of aquatic surface vessels performing an intruder detection task. In our ongoing work, we will continue to study the application of collective sensing in real-robot scenarios.

The concept of collective sensors proposed in this paper opens several new avenues of research. Observations made by different robots can be integrated to allow more precise information to be obtained about the environment. It might be beyond the capability of a single robot to, for instance, estimate the velocity, shape, or size of a particular object, but such estimates could be obtained by combining the sensory readings of multiple robots. Moreover, the sharing of sensory information potentially introduces redundancy in a swarm robotics system. Such redundancy could be used to detect faults, and in case of failure in onboard sensors, a robot could continue to contribute by relying on its collective sensors.

One of the most promising applications of our approach is in the context of heterogeneous swarms [15], where different robots have different capabilities. On the one hand, robots with limited capabilities can benefit from sensory readings shared by robots with different or more capable sensors. On the other hand, by equipping different robots with different sensors, it becomes possible to enhance the capabilities of the swarm as a whole. In ongoing work, we are exploring the application of our approaches to robots with unreliable onboard sensors and to heterogeneous swarms of robots.

Acknowledgements. This work was supported by Fundação para a Ciência e a Tecnologia (FCT) under the grants, SFRH/BD/76438/2011, EXPL/EEI-AUT/0329/2013 and UID/EEA/50008/2013.

References

1. Balch, T., Arkin, R.C.: Communication in reactive multiagent robotic systems. Auton. Robots **1**(1), 27–52 (1994)
2. Bassler, B.L.: How bacteria talk to each other: regulation of gene expression by quorum sensing. Curr. Opin. Microbiol. **2**(6), 582–587 (1999)
3. Beckers, R., Holland, O.E., Deneubourg, J.L.: From local actions to global tasks: stigmergy and collective robotics. In: Proceedings of the International Workshop on the Synthesis and Simulation of Living Systems (ALIFE), pp. 181–189. MIT Press, Cambridge (1994)
4. Beer, R.D., Gallagher, J.C.: Evolving dynamical neural networks for adaptive behavior. Adapt. Behav. **1**(1), 91–122 (1992)
5. Chaimowicz, L., Campos, M.F.M., Kumar, R.V.: Dynamic role assignment for cooperative robots. In: Proceedings of the IEEE International Conference on Robotics and Automation (ICRA), vol. 1, pp. 293–298. IEEE Press, Piscataway (2002)
6. Chandrasekaran, S., Hougen, D.F.: Swarm intelligence for cooperation of bio-nano robots using quorum sensing. In: Proceedings of the Bio Micro and Nanosystems Conference (BMN), p. 104. IEEE Press, Piscataway (2006)
7. Christensen, A.L., O'Grady, R., Dorigo, M.: SWARMORPH-script: a language for arbitrary morphology generation in self-assembling robots. Swarm Intell. **2**(2–4), 143–165 (2008)
8. Christensen, A.L., O'Grady, R., Dorigo, M.: From fireflies to fault tolerant swarms of robots. IEEE Trans. Evol. Comput. **13**(4), 754–766 (2009)
9. Christensen, A.L., Oliveira, S., Postolache, O., de Oliveira, M.J., Sargento, S., Santana, P., Nunes, L., Velez, F., Sebastiao, P., Costa, V., et al.: Design of communication and control for swarms of aquatic surface drones. In: 7th International Conference on Agents and Artificial Intelligence (ICAART). SciTePress, Lisbon (2015)
10. Cianci, C.M., Raemy, X., Pugh, J., Martinoli, A.: Communication in a swarm of miniature robots: the e-puck as an educational tool for swarm robotics. In: Şahin, E., Spears, W.M., Winfield, A.F.T. (eds.) SAB 2006 Ws 2007. LNCS, vol. 4433, pp. 103–115. Springer, Heidelberg (2007)
11. Correll, N., Martinoli, A.: Collective inspection of regular structures using a swarm of miniature robots. In: Ang Jr., M.H., Khatib, O. (eds.) Experimental Robotics IX. STAR, vol. 21, pp. 375–386. Springer, Berlin (2006)
12. Dias, M.B., Zlot, R., Kalra, N., Stentz, A.: Market-based multirobot coordination: a survey and analysis. Proc. IEEE **94**(7), 1257–1270 (2006)
13. Dias, M.B., Ghanem, B., Stentz, A.: Improving cost estimation in market-based coordination of a distributed sensing task. In: Proceedings of the IEEE/RSJ International Conference on Intelligent Robots and Systems (IROS), pp. 3972–3977. IEEE Press, Piscataway (2005)
14. Dietl, M., Gutmann, J.S., Nebel, B.: Cooperative sensing in dynamic environments. In: Proceedings of the IEEE/RSJ International Conference on Intelligent Robots and Systems (IROS), pp. 1706–1713. IEEE Press, Piscataway (2001)

15. Dorigo, M., Floreano, D., Gambardella, L.M., Mondada, F., Nolfi, S., Baaboura, T., Birattari, M., Bonani, M., Brambilla, M., Brutschy, A., Burnier, D., Campo, A., Christensen, A.L., Decugnière, A., Di Caro, G., Ducatelle, F., Ferrante, E., Förster, A., Guzzi, J., Longchamp, V., Magnenat, S., Martinez Gonzales, J., Mathews, N., O'Grady, R., Pinciroli, C., Pini, G., Rétornaz, P., Roberts, J., Sperati, V., Stirling, T., Stranieri, A., Stützle, T., Trianni, V., Tuci, E., Turgut, A.E., Vaussard, F., Montes de Oca, M.: Swarmanoid: a novel concept for the study of heterogeneous robotic swarms. IEEE Robot. Autom. Mag. **20**(4), 60–71 (2013)
16. Dorigo, M., Trianni, V., Şahin, E., Groß, R., Labella, T., Baldassarre, G., Nolfi, S., Deneubourg, J., Mondada, F., Floreano, D., Gambardella, L.M.: Evolving self-organizing behaviors for a swarm-bot. Auton. Robots **17**(2), 223–245 (2004)
17. Duarte, M., Christensen, A.L., Oliveira, S.: Towards artificial evolution of complex behaviors observed in insect colonies. In: Antunes, L., Pinto, H.S. (eds.) EPIA 2011. LNCS, vol. 7026, pp. 153–167. Springer, Heidelberg (2011)
18. Duarte, M., Silva, F., Rodrigues, T., Oliveira, S.M., Christensen, A.L.: JBotEvolver: a versatile simulation platform for evolutionary robotics. In: Proceedings of the International Conference on the Synthesis & Simulation of Living Systems (ALIFE), pp. 210–211. MIT Press, Cambridge (2014)
19. Einolghozati, A., Sardari, M., Fekri, F.: Collective sensing-capacity of bacteria populations. In: Proceedings of the IEEE International Symposium on Information Theory (ISIT), pp. 2959–2963. IEEE Press, Piscataway (2012)
20. Floreano, D., Keller, L.: Evolution of adaptive behaviour in robots by means of Darwinian selection. PLoS Biol. **8**(1), e1000292 (2010)
21. Floreano, D., Mitri, S., Magnenat, S., Keller, L.: Evolutionary conditions for the emergence of communication in robots. Curr. Biol. **17**(6), 514–519 (2007)
22. Fredslund, J., Matarić, M.J.: A general algorithm for robot formations using local sensing and minimal communication. IEEE Trans. Robot. Autom. **18**(5), 837–846 (2002)
23. Ayorkor Korsah, G., Dias, M.B., Stentz, A.: A comprehensive taxonomy for multi-robot task allocation. Int. J. Robot. Res. **32**(12), 1495–1512 (2013)
24. Garnier, S., Jost, C., Gautrais, J., Asadpour, M., Caprari, G., Jeanson, R., Grimal, A., Theraulaz, G.: The embodiment of cockroach aggregation behavior in a group of micro-robots. Artif. Life **14**(4), 387–408 (2008)
25. Gerkey, B.P., Matarić, M.J.: Pusher-watcher: an approach to fault-tolerant tightly-coupled robot coordination. In: Proceedings of IEEE International Conference on Robotics and Automation, (ICRA), pp. 464–469. IEEE Press, Piscataway (2002)
26. Gerkey, B.P., Matarić, M.J.: Sold!: Auction methods for multirobot coordination. IEEE Trans. Robot. Autom. **18**(5), 758–768 (2002)
27. Gutiérrez, A., Campo, A., Dorigo, M., Amor, D., Magdalena, L., Monasterio-Huelin, F.: An open localization and local communication embodied sensor. Sensors **8**(11), 7545–7563 (2008)
28. Hoff, N.R., Sagoff, A., Wood, R.J., Nagpal, R.: Two foraging algorithms for robot swarms using only local communication. In: Proceedings of the IEEE International Conference on Robotics and Biomimetics (ROBIO), pp. 123–130. IEEE Press, Piscataway (2010)
29. Kalman, R.E.: A new approach to linear filtering and prediction problems. J. Fluids Eng. **82**(1), 35–45 (1960)
30. Karol, A., Williams, M.-A.: Distributed sensor fusion for object tracking. In: Bredenfeld, A., Jacoff, A., Noda, I., Takahashi, Y. (eds.) RoboCup 2005. LNCS (LNAI), vol. 4020, pp. 504–511. Springer, Heidelberg (2006)

31. Mamei, M., Zambonelli, F.: Physical deployment of digital pheromones through RFID technology. In: Proceedings of the IEEE Swarm Intelligence Symposium (SIS), pp. 281–288. IEEE Press, Piscataway (2005)
32. Mathews, N., Valentini, G., Christensen, A.L., O'Grady, R., Brutschy, A., Dorigo, M.: Spatially targeted communication in decentralized multirobot systems. Auton. Robots **38**(4), 439–457 (2015)
33. Mondada, F., Bonani, M., Raemy, X., Pugh, J., Cianci, C., Klaptocz, A., Magnenat, S., Zufferey, J.C., Floreano, D., Martinoli, A.: The e-puck, a robot designed for education in engineering. In: Proceedings of the IEEE International Conference on Autonomous Robot Systems and Competitions (ROBOTICA), pp. 59–65. IPCB: Instituto Politécnico de Castelo Branco (2009)
34. Nolfi, S., Floreano, D.: Evolutionary Robotics: The Biology, Intelligence, And Technology of Self-organizing Machines. MIT Press, Cambridge (2000)
35. Pagello, E., D'Angelo, A., Menegatti, E.: Cooperation issues and distributed sensing for multirobot systems. Proc. IEEE **94**(7), 1370–1383 (2006)
36. Parker, L.E.: Alliance: an architecture for fault tolerant multirobot cooperation. IEEE Trans. Robot. Autom. **14**(2), 220–240 (1998)
37. Payton, D.W., Daily, M.J., Hoff, B., Howard, M.D., Lee, C.L.: Pheromone robotics. Auton. Robots **11**(3), 319–324 (2001)
38. Roth, M., Vail, D., Veloso, M.: A real-time world model for multi-robot teams with high-latency communication. In: Proceedings of the IEEE/RSJ International Conference on Intelligent Robots and Systems (IROS), pp. 2494–2499. IEEE Press, Piscataway (2003)
39. Şahin, E.: Swarm robotics: from sources of inspiration to domains of application. In: Şahin, E., Spears, W.M. (eds.) Swarm Robotics 2004. LNCS, vol. 3342, pp. 10–20. Springer, Heidelberg (2005)
40. Schmickl, T., Crailsheim, K.: Trophallaxis within a robotic swarm: bio-inspired communication among robots in a swarm. Auton. Robots **25**(1–2), 171–188 (2008)
41. Stamatis, P.N., Zaharakis, I.D., Kameas, A.D.: A study of bio-inspired communication scheme in swarm robotics. In: Demazeau, Y., Pavón, J., Corchado, J.M., Bajo, J. (eds.) 7th International Conference on Practical Applications of Agents and Multi-Agent Systems (PAMS). AISC, vol. 55, pp. 383–391. Springer, Heidelberg (2009)
42. Stone, P., Veloso, M.: Task decomposition, dynamic role assignment, and low-bandwidth communication for real-time strategic teamwork. Artif. Intell. **110**, 241–273 (1999)
43. Støy, K.: Using situated communication in distributed autonomous mobile robotics. In: Proceedings of the Scandinavian Conference on Artificial Intelligence (SCAI), pp. 44–52. IOS Press, Amsterdam (2001)
44. Stroupe, A.W., Martin, M.C., Balch, T.: Distributed sensor fusion for object position estimation by multi-robot systems. In: Proceedings of the IEEE International Conference on Robotics and Automation, ICRA, pp. 1092–1098. IEEE Press, Piscataway (2001)
45. Svennebring, J., Koenig, S.: Building terrain-covering ant robots: a feasibility study. Auton. Robots **16**(3), 313–332 (2004)
46. Trianni, V., Groß, R., Labella, T.H., Şahin, E., Dorigo, M.: Evolving aggregation behaviors in a swarm of robots. In: Banzhaf, W., Ziegler, J., Christaller, T., Dittrich, P., Kim, J.T. (eds.) ECAL 2003. LNCS (LNAI), vol. 2801, pp. 865–874. Springer, Heidelberg (2003)

47. Turgut, A.E., Çelikkanat, H., Gökçe, F., Şahin, E.: Self-organized flocking with a mobile robot swarm. In: Proceedings of the International Joint Conference on Autonomous Agents and Multiagent Systems (AAMAS), pp. 39–46. IFAAMAS, Richland (2008)
48. Zlot, R., Stentz, A., Dias, M.B., Thayer, S.: Multi-robot exploration controlled by a market economy. In: Proceedings 2002 IEEE International Conference on Robotics and Automation, (ICRA), vol. 3, pp. 3016–3023. IEEE Press, Piscataway (2002)

A Question of Balance
The Benefits of Pattern-Recognition When Solving Problems in a Complex Domain

Martyn Lloyd-Kelly[1]($\boxtimes$), Fernand Gobet[1], and Peter C.R. Lane[2]

[1] Psychological Sciences, University of Liverpool, Liverpool L69 3BX, UK
{M.Lloyd-Kelly,Fernand.Gobet}@liverpool.ac.uk
[2] School of Computer Science, University of Hertfordshire,
Hertfordshire AL10 9AB, UK
peter.lane@bcs.org.uk

Abstract. The dual-process theory of human cognition proposes the existence of two systems for decision-making: a slower, deliberative, *problem-solving* system and a quicker, reactive, *pattern-recognition* system. We alter the balance of these systems in a number of computational simulations using three types of agent equipped with a novel, hybrid, human-like cognitive architecture. These agents are situated in the stochastic, multi-agent *Tileworld* domain, whose complexity can be precisely controlled and widely varied. We explore how agent performance is affected by different balances of problem-solving and pattern-recognition, and conduct a sensitivity analysis upon key pattern-recognition system variables. Results indicate that pattern-recognition improves agent performance by as much as 36.5 % and, if a balance is struck with particular pattern-recognition components to promote pattern-recognition use, performance can be further improved by up to 3.6 %. This research is of interest for studies of expert behaviour in particular, and AI in general.

1   Introduction

The notion of a *dual-process* cognitive system proposes that, given a problem, humans are equipped with two systems to make a decision about what to do. The formal logic-like *problem-solving* system is slow and deliberative, whereas the quicker *pattern-recognition* system uses fuzzier judgements of pattern similarity to propose solutions [40].

The pattern-recognition system has been proposed to be capable of creating, retrieving and using *productions* (a prescribed action for a particular environment state) to achieve the agent's relevant goal(s) [13,40]. Psychological validity of this system is buttressed by human experimental evidence [15,46] and implementations in computational cognitive architectures designed to emulate and explain human cognition [43].

Tension between use of problem-solving and pattern-recognition to solve problems has been identified by many [20,47], resulting in the proposal that domain experts are, in some cases, inflexible problem-solvers, since they are so

© Springer-Verlag Berlin Heidelberg 2015
N.T. Nguyen et al. (Eds.): TCCI XX, LNCS 9420, pp. 224–258, 2015.
DOI: 10.1007/978-3-319-27543-7_11

entrenched in established paradigms [39,41]. This has been proven to be true, but only to a certain degree of expertise; once an above-average level of knowledge has been acquired about a domain, the so-called *Einstellung effect* (a hallmark of expert inflexibility where satisfactory solutions block better ones) is removed [6].

A quantitative, scientific analysis of the potential effects on agent performance by weighting the usage of these systems and configuring their constituent components differently in particular complexities of a stochastic environment is lacking, to our knowledge, in the literature. So, in this paper, we provide such an analysis by investigating what balance of problem-solving and pattern-recognition is most effective when these systems are encapsulated in agents that are equipped with a human-like computational architecture of cognition. These agents are then situated in an environment whose complexity can vary considerably and where precise compile-time prescriptions of optimal actions using techniques such as Markov Decision Processes are implausible.

We compare three ways of making decisions: pure problem solving, "pure" pattern-recognition, and an equal mixture of the two decision-making systems. Our results are especially interesting for those who intend to design robust, self-learning systems that must achieve high levels of performance using information that is learned, whilst being engaged in reactive, sequential decision-making tasks. Our results are also of interest to those who wish to understand how pattern-recognition can improve performance in complex domains and how environment complexity affects the learning rates and performance of autonomous agents in general.

Section 2 discusses the simulation environment in detail and justifies its applicability. Section 3 presents a relevant overview of human cognition, and discusses the computational cognitive architecture in detail. Section 4 covers the implementation details of the agents and Sect. 5 outlines the simulations run to gather data to answer the research questions posed. Section 6 delineates the results obtained and how they have emerged. The paper concludes with Sect. 7, containing the salient points raised by the simulation results, their implications for the current state of the art, and some future research directions.

2   Simulation Environment

The Tileworld environment [33] is a two-dimensional grid of homogeneously-sized squares that contains (along with multiple agents) a number of tiles and holes that exist for a finite period of time (see Fig. 1). An agent can move by up to one square per action, and its main goal is to push tiles into holes, earning the agent a reward and causing the tile and hole in question to disappear. An agent's actions and goal achievement are episodic and delayed, since several actions may need to be performed in succession before a hole is filled, and only one tile can be pushed at any time by an agent. Explicit obstacles have not been included in this version of Tileworld, since tiles, holes and agents act as natural obstacles, given that a square can only be occupied by one object at a time.

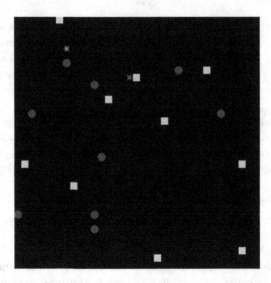

Fig. 1. Tileworld environment: agents are denoted by red and green turtle shapes, tiles by yellow squares and holes by blue circles (Color figure online).

Environmental complexity can be controlled by altering parameter values to manipulate *intrinsic* and *extrinsic* complexity. Intrinsic complexity is complexity that the environment has direct control over whereas extrinsic complexity is complexity that the environment has no direct control over. In our simulations, intrinsic complexity is controlled by parameters that define when new tiles and holes can be created, the probability of a new tile or hole being created, and how long these artefacts exist for before being removed. Extrinsic complexity is controlled solely by altering the number of agents present in the environment: introducing more agents increases the environment state space by virtue of extra agents but also by increasing the chance of environment resources being interacted with. This potentially produces more environment states and increases environment complexity.

In a simplified version of Tileworld where only holes and one agent (no opponents) may exist, optimal policy calculations for a computer with reasonable resources using a Markov Decision Process (MDP) becomes intractable when the total number of squares in the Tileworld equals 16 or 25 (a 4×4 or 5×5 grid) [37]. In comparison, the Tileworld implemented in this paper is much more complex: tiles, holes, and opponents can exist simultaneously in mixed quantities and the environment comprises 1225 (35×35) squares in total (however, an agent can only "see" 25 squares in total at a time; see Sect. 4).

Taking the complexity of the entire environment into account at once, an environment with 2 to 8 agents yields $\approx 10^{589}$ to $\approx 10^{600}$ possible states (these numbers cover the range of agents used, see Sect. 5). From the perspective of a single agent, complexity ranges from $\approx 10^{12}$ states (2 agents) to $\approx 10^{14}$ states (8 agents).

Consequently, we assert that the complexity of this environment, along with the ability to exert fine-grained control over this complexity, provides a suitable test-bed to address the research questions posed.

3   Cognitive Architecture

In studying human cognition, much scrutiny has been focused upon explaining expert behaviour in complex domains; chess, in particular, has benefited from such effort. Research has identified that the difference in performance between chess masters and amateurs hinges upon the breadth and quality of knowledge possessed by masters [11].

The total amount of information in chess has been calculated to contain 143.09 bits of information or, 10^{43} positions ($2^{143.09}$) [23]. However, some of these positions are redundant or implausible; rectified calculations give a total space of 50 bits of information or, 10^{15} positions [11]. A promising psychological theory that accounts for the ability of chess masters to learn and retain such large[1] amounts of information, given known limitations on human cognition, is *chunking theory* [31]. Chunking theory suggests that information in memory is stored as aggregated environmental information or *chunks*; their existence and application in chess has been established through rigorous testing (see [8,11] for good examples). Indeed, computational cognitive models that implement chunking theory have closely mimicked human behaviour in domains other than chess too (see Sect. 3.1 for details).

With regard to decision-making, chess masters demonstrate pattern-recognition frequently: key features of certain board configurations are recognised extremely quickly [10] and typical, good moves are used when better, relatively uncommon moves exist [6,35]. This indicates that when an adequate knowledge base exists for a domain, pattern-recognition is the preferred *modus operandi* for human decision-making (examples in other domains are given in [6]). However, given that there may exist many possible solutions for a particular situation, how exactly is a decision regarding what to do reached?

One proposal is that pattern-recognition is underpinned by the existence and use of *productions* stored in memory. We define productions as being *if-else* conditions where the *if* component is visual information and the *then* component is an action that should be performed. Productions can be assigned utility ratings by a process akin to reinforcement learning [44], whose presence in human cognition has been extensively validated [12,21]. These ratings enable production selection to be *rational*: better productions should be selected more frequently than worse ones [32]. When applied to domains like Tileworld, where an agent's actions and goal achievement are episodic, rating production utility entails using *discounted rewards*: productions performed closer to the time a reward for goal achievement is received are assigned, or have their utility ratings incremented by, an amount that reflects a greater share of the reward received than productions

---

[1] With respect to both the number of positions and the amount of information within each position.

performed further in the past, according to a discount factor β, $(0 < \beta < 1)$ [19]. Human production selection is also non-deterministic; worse productions may be selected given better alternatives [9,34].

Consequently, we have used the *Chunk Hierarchy and REtrieval STructures* (CHREST) architecture that implements chunking theory computationally [11,16] as the core of our cognitive architecture. Additionally, we have implemented a pattern-recognition and problem-solving system. The pattern-recognition system is considered a part of CHREST since it depends upon particular CHREST components and handles CHREST-compatible information. The problem-solving system is domain-specific and does not directly interact with any part of CHREST.

To enable correct operation of the pattern-recognition system, the *Profit Sharing with Discount Rate* (PSDR) reinforcement learning theory [4] and the *Roulette* selection algorithm [5] have also been used. The CLARION cognitive architecture [42,43] adopts a similar combination of elements but usually combines a backpropagation network, Q-learning theory [45] and a Boltzmann distribution selection algorithm. However, such a combination is not capable of handling stochastic environments so is unsuitable for our purposes; Q-learning in particular fails to converge in domains similar to and smaller than the version of Tileworld used in this paper [4]. PSDR also requires that an agent have access to an *episodic memory* data structure that tracks what actions have been performed by the agent in response to particular environmental stimuli.

Due consideration must be afforded to how the problem-solving and pattern-recognition systems exert control over decision-making given that we are interested in examining how a human-like system of cognition performs in the stochastic domain described. The research discussed proposes that problem-solving and pattern-recognition never operate simultaneously in human cognition. So, after observing the current state of the environment, agents will first use pattern-recognition to generate a solution. If no solution is output from the pattern-recognition system, the problem-solving system is used instead. Implementing decision-making in this way creates a *modular* dual-process decision-making system [26], and extends CHREST's existing functionality.

In this section, Sect. 3.1 discusses pertinent features of CHREST and Sects. 3.2 and 3.3 outline the problem-solving and pattern-recognition systems. Section 3.4 details the episodic memory structure and Sects. 3.5 and 3.6 discuss the PSDR and Roulette algorithms.

3.1   CHREST

CHREST is an example of a symbolic cognitive computational architecture [36] and is capable of creating extensible knowledge-bases that enable human-like storage, organisation and retrieval of memory. CHREST also provides functionality to create productions by generating directed links between chunks and can store utility ratings for productions by associating numeric values with these links. The validity of CHREST as a theory of human-like cognition has been established in a variety of domains, including board games [7,11], implicit learning [26] and natural language acquisition [14,22]. A version of CHREST similar to that described in this section has also investigated how binding rationality affects the performance and learning rates of Tileworld agents [29].

CHREST comprises two main components: *short-term memory* (STM) and *long-term memory* (LTM). Unlike cognitive architectures such as Soar [25] and ACT-R [2], CHREST does not encode information in LTM as procedural, declarative or semantic. Rather, information is organised according to its *modality*, of which CHREST defines three types: action, visual and verbal (no verbal information is used in our simulations).[2]

Knowledge Representation. In the CHREST framework, units of information called *patterns* are used to represent knowledge about the environment and chunks refer to LTM knowledge. To create a chunk, a single pattern or combination of patterns must be used to create or modify a LTM *node*. When this occurs, the input pattern becomes, or is added to, a node's *image*. Therefore, a chunk is strictly defined as being the content of a node's image.

Since patterns represent knowledge about the environment, they are created by an agent's domain-specific *input/output* component (see Sect. 4.2 for implementation details of this component). In the simulations run, the patterns used are instances of three-item tuples called *item-on-square* patterns that contain an identifier string and two numbers (see captions for Figs. 2 and 3 for examples).

For visual item-on-square patterns, the identifier string represents the *class* of object seen. In these simulations, T encodes a tile, H encodes a hole and A encodes another agent. An agent does not encode its own location in a pattern (and consequently, the chunks it creates) since all objects are encoded relative to its current location. The first and second numbers represent how many squares to the north and east of the agent the object is, respectively.[3] For example, the visual pattern <[T 1 2]>, states that there is a tile (T) located 1 square north (1) and 2 squares east (2) of the agent's current location.

For action item-on-square patterns, the identifier string represents an action that an agent should perform (see Table 1 for the mappings between identifiers and actions). The first of the two numbers represents the compass direction the agent should face when performing the action: 0 = north, 90 = east etc., and the second number denotes the number of squares that should be moved by the agent when the action is performed; equal to either 0 (for the *remain-stationary* and *problem-solve* actions only) or 1 (see Sect. 2). For example, the action pattern <[PT 90 1]> states that an agent should face east (90) and perform the *push-tile* action (PT), pushing a tile and moving itself 1 square (1) in this direction. There are a total of 18 actions that can be performed by an agent: 4 variations of 4 actions plus 2 actions with no variations.

---

[2] See [28] for a detailed comparison of ACT-R and CHREST's LTM implementation.
[3] South and west are represented by negative numbers.

Table 1. Mappings of identifiers for action item-on-square patterns, the action identified and variations thereof.

Identifier	Action	Variations
MR	Move-randomly	North, east, south, west
MAT	Move-around-tile	
MTT	Move-to-tile	
PT	Push-tile	
RS	Remain-stationary	N/A
PS	Problem-solve	

Short and Long-Term Memory. STM consists of three fixed-length, first-in-first-out lists, one for each modality supported in CHREST. These lists store chunks that have been retrieved from LTM.

LTM is composed of a discrimination network that acts as a retrieval device and a similarity function; it is analogous to the hidden layers of a connectionist network, or the RETE network of Soar [25]. Nodes in LTM are connected using *test links*; to learn or retrieve information from LTM, a pattern, ϕ, is input to LTM and the discrimination network is traversed by sorting ϕ from the general root node of LTM to the corresponding modality root node and then along matching test links until a leaf node is reached or no further test links match. If ϕ does not match the chunk retrieved after traversal, θ, the network is modified using one of two learning mechanisms: *discrimination* or *familiarisation*. Implementation details of these mechanisms that are of interest to our simulations are discussed in the following section.[4]

Discrimination, Familiarisation and Production Creation. These mechanisms take a user-defined period of time; performing one blocks performance of another. Hence, the model of cognition implemented is very human-like: learning is slow when an agent is first placed in an environment but accelerates as interaction occurs.

Discrimination. Increases the number of chunks stored in LTM and occurs when a pattern input to LTM, ρ (this may be a sub-pattern of a larger pattern, ϕ), is either not present in LTM or is present in LTM but the chunk retrieved, θ, does not contain ρ, and θ is "finished": no new patterns can be appended to it (indicated by a dollar sign, $). In the first case, a new node is created and connected to the relevant modality root node (see Fig. 2(b)). In the second case, a new node is created and connected to θ (see Fig. 2(c)). The connection created is a test-link that contains ρ.

Familiarisation. Increases the size of a chunk and occurs when a chunk, θ, is retrieved from LTM given a pattern, ϕ, as input and the following conditions are all true:

---

[4] See [18,26] for further details of these mechanisms.

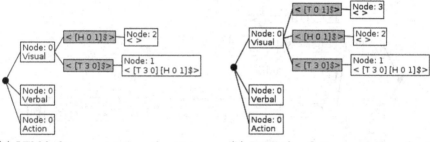

(a) LTM before presentation of pattern.

(b) LTM after first presentation of pattern.

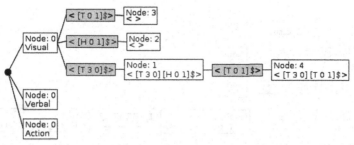

(c) LTM after second presentation of pattern.

Fig. 2. Discriminating visual item-on-square pattern <[T 3 0][T 0 1]$> that indicates locations of two tiles relative to an agent's location: first is 3 squares north, second is 1 square east. Test links are indicated by grey rectangles.

- ϕ contains a sub-pattern, ρ, that exists as a chunk in LTM.
- ρ is not present in θ.
- Sub-patterns preceding ρ in θ and ϕ are the same.
- θ is not "finished".

If the above conditions are all true, CHREST adds ρ to θ (see Fig. 3).

Production Creation. Productions are created from information contained in *episodes* (see Sect. 3.4) and implemented using hash map data structures contained in visual LTM nodes; keys contain pointers to action LTM nodes and values denote the production's utility rating, as defined in the introduction to Sect. 3 and illustrated in Fig. 4. Attempts to create productions that already exist are ignored.

For a production to be created, the visual and action parts of the production must be capable of being *recognised*, i.e. a chunk must be returned when each part is input to LTM. Due to CHREST's intention to simulate human cognition as closely as possible, it may be that, after passing the visual part of a production, ϕ, as input to LTM, the recognised visual chunk, θ, does not match ϕ exactly. Instead, θ may only contain *some* sub-patterns common to itself and ϕ. For example, if the visual pattern <[T 1 0][H 2 0]$> is passed to LTM as input, <[T 1 0]> may be retrieved if <[T 1 0][H 2 0]$> has not been fully

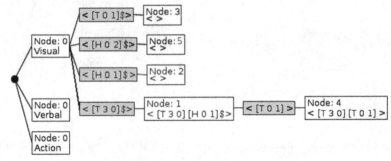

(a) LTM before presentation of pattern.

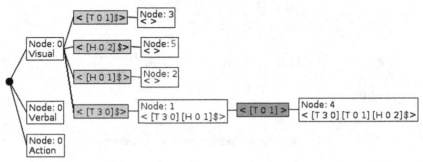

(b) LTM after presentation of pattern.

Fig. 3. Familiarising visual item-on-square pattern <[T 3 0][T 0 1][H 0 2]$> that indicates locations of two tiles and one hole relative to an agent's location: first tile is 3 squares north, second is 1 square east and hole is 2 squares east.

familiarised. Thus, *over-generalisation* of production selection may occur, a very human-like cognition trait [24].

In these simulations, two broad types of production can exist in LTM and are differentiated by the type of action node they terminate with. The first production type terminates at an explicit action node, for example: *push-tile north*, the second terminates at an action node that prescribes use of the problem-solving system. Differences in how these two production types are handled embody the three types of decision-making mentioned in Sect. 1 and create the three agent types discussed in Sect. 4.1.

3.2   Problem-Solving System

Actions generated by the problem-solving system are intended to achieve the agent's currently active goal; these are not explicitly represented in any data structure available to the agent but are ostensibly activated after analysing visual input from the environment. The result of this analysis is used to run one of three hand-coded procedures: **move randomly**, **secure tile** or **push tile to hole**. Note that we have conflated the concepts of *goal* and *environment state* since these have a simple one-to-one mapping in the environment modelled.

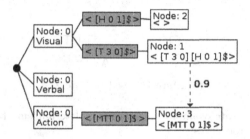

Fig. 4. An example production with an utility rating of 0.9. The production reads as: "If I can see a tile 3 squares north and a hole 1 square east then, I should move north to the tile."

There are three sub-goals that need to be achieved to fulfil the agent's main goal of `fill hole with tile`. These are: `find tile`, `secure tile` and `find hole`. The problem-solving system therefore follows the procedure outlined below. Active goals are highlighted using `fixed-width` font, procedures run are highlighted in **bold** and actions generated are highlighted in *italics*. Note that "adjacent" is defined as an object being one square north, east, south or west of the object referred to.

1. Agent is surrounded, i.e. all adjacent squares are occupied by non-movable tiles, holes or other agents: *remain stationary* generated.
2. Agent is not surrounded.
   – Tiles and holes can be seen: determine closest hole to the agent, H, and tile that is closest to H, T.
     • T is adjacent to agent and can be pushed closer to H from agent's current position: `fill hole with tile` activated, **push tile to hole** run, *push tile* generated.
     • T is adjacent to agent but cannot be pushed closer to H from agent's current position: `secure tile` activated, **secure tile** run, *move around tile* generated.
     • T is not adjacent to agent: `secure tile` activated, **secure tile** run, *move to tile* generated.
   – Tiles can be seen but no holes: determine distance of agent from closest tile, T.
     • T is adjacent to agent: `find hole` activated, **push tile to hole** run, *push tile* generated.
     • T is not adjacent to agent: `secure tile` activated, **secure tile** run, *move to tile* generated.
   – No tiles can be seen: `find tile` activated, **move randomly** run, *move randomly* generated.

Note that some procedures generate actions non-deterministically in some circumstances; consider the environment states in Fig. 5. The active goal of agent A in both states is `secure tile`, specifically, tile $T1$, so it runs the **secure tile** procedure to generate an action to try and achieve this goal. The optimal action

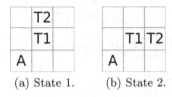

(a) State 1.   (b) State 2.

Fig. 5. Environment state examples to justify non-determinism of action production by problem-solving procedures.

in the case of Fig. 5(a) is for A to move north around $T1$ so that it is able to push $T1$ to the east thus securing it. However, this action is non-optimal if the environment state in Fig. 5(b) is considered since A cannot push $T1$ east because $T2$ blocks $T1$ along this heading. Consequently, the optimal action in one state may be the non-optimal action in a similar state. So, in this case, the secure tile procedure has a 0.5 probability of generating either a *move around tile north* or a *move around tile east* action.

3.3 Pattern-Recognition System

The pattern-recognition system's operation is intended to be analogous to habitual behaviour in human beings: behaviours become habitual when they are frequently selected in response to particular goals being activated [1]. The system therefore uses visual patterns and production utility ratings as input to propose actions to perform. Other than the fact that the pattern-recognition system is considered to be a part of CHREST whereas the problem-solving system is not (see introduction to Sect. 3), there are two crucial differences between this system and the problem-solving system:

1. The pattern-recognition system cannot generate novel actions, it can only select actions contained in existing productions.
2. The pattern-recognition system may have to choose an action from many potential productions, depending upon how many productions exist for the visual pattern input to LTM.

After inputting a visual pattern, ϕ, to the pattern-recognition system, an attempt is made to recognise and retrieve productions from LTM, Ψ. If no productions are retrieved, execution of the system halts and decision-making control is passed to the problem-solving system. Otherwise, the utility ratings of Ψ are used as input to the Roulette selection algorithm (see Sect. 3.6) to select an action for execution.

If a production is selected by the pattern-recognition system, its action is not passed as input to CHREST to be learned if the action is performed successfully (see Sect. 4.3). This is because the action must have been learned for CHREST to have created the production (see Sect. 3.1: *Discrimination, Familiarisation and Production Creation*). Therefore, further learning of the action is redundant.

3.4   Episodic Memory

The episodic memory structure used by agents is analogous to STM and is therefore implemented as a fixed-length first-in-first-out list. An episode contains four pieces of information: a visual pattern, v, an action pattern, α (executed by the agent in response to v), the time α was executed (required by PSDR, see Sect. 3.5) and whether α was produced by the problem-solving or pattern-recognition system (enables type 3 agents to modify productions correctly, see Sect. 4.3).

3.5   Profit Sharing with Discount Rate

PSDR uses a *credit assignment function* (1) to calculate production utility ratings, P_σ. For example: at time t, an agent executes an action in response to the current visual environment state generating a episode, E_t. At time $t + 1$, the agent executes another action in response to the current visual environment state, producing another episode, E_{t+1}, and continues this cycle until it receives a reward, R, at time T. At time T, the agent's episodic memory will contain the following episodes if the number of episodes from E_t to E_T is less than, or equal to, the maximum size of episodic memory: $(E_t, E_{t+1} \ldots E_T)$. With $R = 1$ and discount rate $\beta = 0.5$, the production corresponding to episode E_T receives 1 as credit, E_{T-1}'s production receives 0.5, E_{T-2}'s production' receives 0.25 etc. The credit generated for a production is then added to that production's current utility rating.

$$P_\sigma = P_\sigma + (R \cdot \beta^{T-t}) \ (0 < \beta < 1) \tag{1}$$

PSDR [4] was chosen as a reinforcement learning theory for three reasons: first, it can be used in domains where mathematical modelling of the domain is intractable, a property of the version of Tileworld implemented (see Sect. 2). Second, PSDR's production utility rating mechanism is congruent with that discussed earlier (see introduction to Sect. 3) since it uses discounted rewards. Third, PSDR's effectiveness in enabling agents to learn and apply robust, effective productions autonomously in dynamic, multi-agent domains that are similar to the version of Tileworld used in these simulations has been validated by others [3,4].

3.6   Roulette Algorithm

The Roulette algorithm [5] uses production utility ratings to select an action for execution given a number of candidate productions. Equation (2), generates a value, ω, for a candidate production, P, from P's utility rating, P_σ, divided by the sum of each candidate production's utility rating, P_σ^n to P_σ^N. Candidate productions are then organised in ascending order according to their ω value and used to create a range of values; candidate productions with greater ω values occupy greater ranges. Finally, a random number, $0 < R < 1$, is generated and used to select a production from the candidates. Therefore, the algorithm is non-deterministic and abides by the principle of rationality defined in the introduction to Sect. 3: candidate productions with greater utility ratings will be more likely to be selected whereas it is less likely for candidate productions with lower utility ratings to be selected.

$$\omega = P_\sigma / \sum_{n=1}^{N} P_\sigma^n \qquad (2)$$

4   Agent Implementation

Agents are equipped with the cognitive architecture described in Sect. 3 and a domain-specific input/output component. Agents are goal-driven, non-communicative, non-cooperative, and have limited vision. The size of an agent's observable environment is controlled by a parameter that takes a number as input to indicate how many squares north, east, south and west the agent can "see". We keep the value of this parameter constant at 2 since agent performance should not be affected by differences in "physical" capabilities. The size of visual patterns generated is dependent on this parameter, so any visual pattern constructed can only contain 24 item-on-square patterns at most.[5] This is important since larger values may result in the agent constantly discriminating and familiarising due to large input patterns and thus blocking production creation (see Sect. 3.1). Setting the sight parameter to 2 is also the minimum value that allows the agent to see "around" a tile so that its ability to be pushed can be determined (important to enable valid solutions to be provided by the problem-solving system).

Figure 6 illustrates the cognitive architecture structures discussed in Sect. 3, the agent-specific components discussed in this section and how information flows between them. Note that the sequencing illustrated in Fig. 6 does not always apply due to the agent's type and how the problem-solving and pattern-recognition systems operate. These details, along with the sequencing changes mentioned, are delineated in Sect. 4.1.

This section proceeds as follows: we discuss the implementation details of the three agent types that embody the three different types of decision-making outlined in Sect. 1 in Sect. 4.1. Operation of the agent-specific input/output component is outlined in Sect. 4.2 and the execution cycle for agents is provided in Sect. 4.3.

4.1   Agent Types

In Sect. 3.1: *Discrimination, Familiarisation and Production Creation*, we delineated two types of productions that can be created by agents: productions terminating with explicit actions, i.e. *push-tile north*, and those terminating with the action that prescribes usage of the problem-solving system. The three types of agents implemented in these simulations are defined by the types of production they can create:

---

[5] A square in Tileworld can only contain one item (see Sect. 2), one item-on-square pattern encodes one item and the agent doesn't encode its own location.

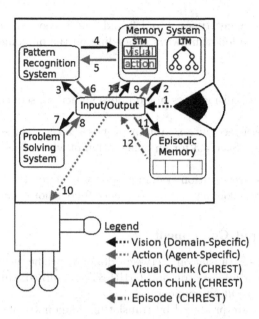

Fig. 6. Agent architecture and flow of information throughout (numbers denote sequence of information processing and are discussed in Sect. 4.1).

- Agent Type 1 (pure problem-solver): neither type of production are created in LTM; agents of this type will always use problem-solving to decide upon what action to perform next.
- Agent Type 2 ("pure" pattern-recogniser): only creates productions that terminate with explicit action chunks.
- Agent Type 3 (problem-solver and pattern-recogniser): creates both types of production.

Since agent type 1 does not use the pattern-recognition system at all, sequences 3–6 and 11–13 do not occur in Fig. 6.

Agent type 2 uses problem-solving to generate actions initially but after a production, P, has been constructed and rated, problem-solving will no longer be used to generate an action when P's visual condition is encountered (pattern-recognition is always used before problem-solving by agents, see introduction to Sect. 3). Hence, agents of type 2 are not "pure" pattern-recognisers (hence the quotes around "pure") in the same sense as agents of type 1 are pure problem-solvers since problem-solving is still used to some degree.

Agent type 3 strikes more of a balance between problem-solving and pattern-recognition system use than agent type 2. The problem-solving system will be used more in initial decision-making (as it is for agent type 2) but, as LTM develops, it may be that productions generated result in either an explicit action being performed or the problem-solving system being used to generate a potentially novel and better action.

With regard to agent types 2 and 3 and Fig. 6, the sequencing illustrated can be altered in the following ways depending on the outcome of the pattern-recognition system:

– Agent types 2 and 3.
  • If pattern-recognition does not propose an action, sequences 5 and 6 not not occur.
  • If the pattern-recognition system proposes an action, sequences 7–9 do not occur (see Sect. 3.3 for an explanation of why 9 does not occur).
– Agent type 3.
  • If the pattern-recognition system proposes an action that prescribes usage of the problem-solving system, sequence 9 does not occur.

4.2 Input/Output Component

In the simulations, translation between domain-specific visual information, agent-specific action information and CHREST is required: this is provided by an agent's input/output component.

Visual patterns are produced by translating domain-specific visual information and output is sent to the agent's memory system to be learned and the pattern-recognition system, if applicable (see Sect. 4.1). The unmodified domain-specific visual information may also be sent to the problem-solving system, if applicable (again, see Sect. 4.1).

Actions produced by the problem-solving or pattern-recognition system are also passed to this component. Agent-specific actions received from the problem-solving system are converted into CHREST-compatible action patterns whereas output from the pattern-recognition system is converted from CHREST-compatible action chunks into agent-specific actions (see Sects. 3.2 and 3.3 for details). In either case, the original action information and its converted form are retained; the CHREST-compatible action pattern is sent to CHREST's memory system so it can be learned whilst the agent-specific version is executed by the agent, causing the agent to perform the action in the environment.

Episodes (see Sect. 3.4) are also encoded and decoded by an agent's input/output component. An example of an episode is as follows: [<[T 1 0] [H 2 0] $> <[PT 0 1] $> 110 true]. This episode should be interpreted thus: an agent saw a tile 1 square to the north and a hole 2 squares to the north (<[T 1 0] [H 2 0] $>) and used its problem-solving system (true) to generate a *push tile north* action (<[PT 0 1] $>) that was executed at time 110 (110). When decoding an episode, the visual and action pattern are retrieved and sent to the agent's memory system to enable production creation or modification (see Sect. 4.3).

4.3 Execution Cycle

The agent execution cycle runs for each agent in turn after every time increment in the Tileworld environment. The order of agent execution is randomised so, at time t, agent 0 may execute first then agent 1 whereas at time $t + 1$, agent 1 may execute before agent 0.

Note that agents have a specific *intention reconsideration* strategy implemented: when the current time, T, equals the time that an action, α, is to be performed, t, the agent generates a new visual pattern, χ, and compares this to the visual pattern, ϕ, used to generate α. If $\phi \neq \chi$, the agent does not perform α and instead generates and loads a new action for execution based upon the information in χ.

The execution cycle proceeds as follows. Note that some steps in the execution cycle require additional explanation to justify their inclusion, these explanations follow the execution cycle delineation. Each agent begins by checking to see if there is an action loaded for execution:

1. No action loaded for execution.
   (a) Generate visual pattern, ϕ.
   (b) Pass ϕ as input to LTM and attempt to learn.
   (c) Use ϕ to generate a new action pattern, α, using problem-solving or pattern-recognition system, depending upon agent type.
   (d) Load α for execution, pass α as input to LTM and attempt to learn. Set time for execution of α to t and exit execution cycle.
2. Action α is loaded for execution, check to see if current time, T, equals t.
   (a) $T = t$: generate new visual pattern, χ, and compare this to ϕ.
      i. $\phi = \chi$: attempt to perform α.
         A. Agent successfully performs α. Check agent type.
            – Agent type 1: exit execution cycle.
            – Agent type 2/3:
               If α is not a *move-randomly* action, create new episode in episodic memory and attempt to create a production in LTM between ϕ and α.
               If α achieves agent's primary goal, apply PSDR to productions representing episodes in episodic memory and clear episodic memory.
         B. Agent unsuccessfully performs α: exit execution cycle.
      ii. $\phi \neq \chi$: unload α for execution and exit execution cycle.
   (b) $T \neq t$: exit execution cycle.

The refusal to create a new episode and production using ϕ and α when α is a *move-randomly* action is due to two reasons. First, if such productions were created, agents would be biased in their random movement so, if the agent selects productions like this in future, the movement performed is not truly random. Second, given the stochastic nature of Tileworld, biasing random movement could artificially influence an agent's performance. For example, it may be that tiles and holes are generated, by chance, to the west of an agent more frequently. If this agent's random movement were biased to the east, it would encounter resources less frequently, reducing its ability to score and impinging its performance.

To justify why agents create/modify productions after an action is successfully performed rather than before, we appeal to the definition of rationality provided in the introduction to Sect. 3: less useful productions should be more strongly suppressed than more useful ones. Actions are only ever not performed because the decision-making system has failed to take into account some environmental resource that stops the action being performed successfully (trying to push a tile along a heading when there is another tile in the way). Therefore, productions that are completely useless are never created.

On this note, it is worthwhile to explain under what circumstances an agent may fail to perform an action since it appears that the intention reconsideration strategy implemented should prevent such an event. Essentially, this can only occur for agents that use pattern-recognition. For example, an agent may attempt to perform a *push-tile* action on tile $T1$ after using pattern-recognition. However, due to over-generalisation of production selection (see Sect. 3.1: *Discrimination, Familiarisation and Production Creation*), it may be that the agent fails to consider a tile present in the visual input that blocks $T1$ from being pushed in the direction suggested. Thus, when the agent attempts to push $T1$, $T1$ is blocked and the action fails.

When creating/modifying production utility ratings after an action is performed, special consideration must be afforded to type 3 agents. Since these agents can create/modify productions whose actions propose using an explicit action or the problem-solving system, they must make a choice given that production creation/modification occurs in CHREST and CHREST is not able to create/modify two productions simultaneously. So, if an episode indicates that its action was generated using problem-solving, type 3 agents create or select the type of production to create/modify randomly. This is implemented by the agent generating a random float R, $(0 <= R < 1)$. If $R < 0.5$ a/the production prescribing use of the problem-solving system given the visual part of the episode is created/modified. Otherwise, a/the production prescribing use of the explicit action performed given the visual part of the episode is created/modified.

5   Simulation Details

Two sets of simulations were run. The first explores what balance of problem-solving and pattern-recognition system use maximises agent performance given different environmental complexities and how differing environment complexities affect pattern-recognition and problem-solving use. The second comprises a sensitivity analysis of episodic memory size and discount rate for agent types 2 and 3 and allows us to ascertain if significantly altering these values affects the dependent variables outlined for the first set of simulations. Hence, this section is split in two: Sect. 5.1 details the first set of simulations, Sect. 5.2 outlines the second.

5.1   Decision-Making System Use, Environment Complexity and Performance Details

For this set of simulations, 27 conditions were simulated and run. Conditions are representative of various degrees of intrinsic/extrinsic environmental complexity and agent types (see Sect. 4.1). Each condition was repeated 10 times to harvest a data set large enough to provide a robust analysis. For each repeat, average frequencies of problem-solving and pattern-recognition system use were recorded along with the average score for all agents (to determine performance). The overall values for these dependent variables over each condition were then calculated by averaging the averages obtained for the repeats.

Our null hypotheses state that:

- Using problem-solving instead of pattern-recognition and vice-versa does not have any significant effect on the performance of agents.
- Altering extrinsic and intrinsic environment complexity does not have any significant effect upon problem-solving or pattern-recognition use.
- Altering extrinsic and intrinsic environment complexity does not have any significant effect upon the performance of agents.

Intrinsic environment complexity is controlled by the values of the *hole appearance probability*, *tile appearance probability*, *hole lifespan* and *tile lifespan* parameters. Greater *tile/hole appearance probability* values and smaller *tile/hole lifespan* values equate to greater complexity; more tiles/holes appear but for shorter periods of time resulting in a greater number of novel environment states occurring. One may expect the values for the *tile/hole appearance interval* parameters to also be varied. However, the intrinsic complexity of the environment can be significantly modified by varying the values of the parameters mentioned. Values for the *tile/hole appearance probability* parameters were derived by simply taking the median probability, 0.5, as the moderate complexity value and then taking the lowest/highest values possible without guaranteeing tile/hole appearance since this would significantly skew the results. The *tile/hole appearance probability* and *tile/hole lifespan* parameter value mappings for each level of environment complexity are provided below:

- Environment complexity: low
  - Tile/hole appearance probability: 0.1
  - Tile/hole lifespan: 80 s
- Environment complexity: moderate
  - Tile/hole appearance probability: 0.5
  - Tile/hole lifespan: 40 s
- Environment complexity: high
  - Tile/hole appearance probability: 0.9
  - Tile/hole lifespan: 20 s

Extrinsic environmental complexity is determined by the number of agents in the environment (see Sect. 2 for justification). We set this variable to either 2, 4 or 8.

All other variable values are kept constant (see Table 2). There are three major groups of conditions differentiated by the degree of intrinsic environment complexity used. These major groups then consist of a further three sub-groups of conditions differentiated by the degree of extrinsic environmental complexity. Finally, each sub-group consists of three sub-sub-groups differentiated by agent type (see Sect. 4.1). Note that the agent types used in a condition are homogeneous.

Table 2. Mappings of independent variable names to owner (agent, CHREST, environment), value used and justification for value used.

Independent variable	Owner	Value	Justification
Problem-solving time	Agent	1 s	Equals value of the tile/hole birth interval parameters so planned actions may be reconsidered due to the appearance of a new tile or hole
Sight radius	Agent	2	See Sect. 4
Add link time	CHREST	10 s	Taken from [38]
Discount rate	CHREST	0.1 to 0.9	Kept constant at median value (0.5) for simulations detailed in Sects. 5.1 and 6.1, varied in simulations detailed in Sects. 5.2 and 6.2
Discrimination time	CHREST	10 s	Taken from [38]
Episodic memory size	CHREST	6 to 14	Kept constant at median value (10) for simulations detailed in Sects. 5.1 and 6.1, varied in simulations detailed in Sects. 5.2 and 6.2
Familiarisation time	CHREST	2 s	Taken from [38]
Pattern-recognition time	CHREST	0.2 s	Taken from [17]
Hole appearance interval	Env.	1 s	Equals value of the problem-solving time parameter so planned actions may be reconsidered due to the appearance of a new tile or hole
Play time	Env.	28800 s	Allows pattern-recognition systems to learn enough information to be useful
Reward value	Env.	1	Equal to the single point received for an agent achieving its main goal of pushing a tile into a hole
Tile appearance interval	Env.	1 s	Equals value of the problem-solving time parameter so planned actions may be reconsidered due to the appearance of a new tile or hole
Time increment	Env.	0.1 s	CHREST operations measured in milliseconds: this is the smallest major time unit possible

5.2 Sensitivity Analysis

In this set of simulations, 270 conditions were simulated and run and each condition was repeated 10 times to harvest a data set large enough to provide a robust analysis. The dependent variables recorded and methods used to calculate their values remain unchanged from the simulations detailed in Sect. 5.1.

Our null hypotheses states that there is no significant effect upon agent performance or pattern-recognition/problem-solving system use when episodic memory size or discount rates are altered.

Note that in this set of simulations, we have conflated extrinsic and intrinsic environment complexity into a generic environment complexity to make the analysis simpler. The *tile/hole appearance probability* and *tile/hole lifespan* parameter value mappings for each level of environment complexity are provided below:

- Environment complexity: low
  - Tile/hole birth probability: 0.1
  - Tile/hole lifespan: 80 s
  - Number of agents: 2
- Environment complexity: moderate
  - Tile/hole birth probability: 0.5
  - Tile/hole lifespan: 40 s
  - Number of agents: 4
- Environment complexity: high
  - Tile/hole birth probability: 0.9
  - Tile/hole lifespan: 20 s
  - Number of agents: 8

We take the values used for the discount rate and episodic memory size in the simulations detailed in Sect. 5.1 as median values for these variables in this set of simulations. Discount rate is then incremented in steps of 0.1 from 0.1 to 0.9 and episodic memory size is incremented in steps of 2 from 6 to 14 for each agent type in each of the environment complexities defined. All other variable values are kept constant (see Table 2).

There are three major groups of conditions differentiated by the degree of environment complexity used. These major groups then consist of a further two groups of conditions differentiated by agent type (agent type in each condition is homogeneous). Each of these agent type groups then consists of five groups determined by size of episodic memory. Finally, each of these episodic memory size groups is divided into nine groups distinguished by the discount rate used. Thus, all combinations of environment complexity, agent type, episodic memory size and discount rates outlined for this set of simulations are tested.

6   Results and Discussion

This section is split into two, with Sect. 6.1 discussing results from the simulations described in Sects. 5.1 and 6.2 discussing results from the simulations described in Sect. 5.2.

6.1   Decision-Making System Use, Performance and Environment Complexity Results

Results in this section were analysed using a $3 \times 3 \times 3$ analysis of variance (ANOVA), with intrinsic complexity, extrinsic complexity and agent type as between-subject variables.[6] As mentioned in Sect. 5.1, we have collected data for three dependent variables: average score, average frequency of problem-solving system use and average frequency of pattern-recognition system use. This section is in three parts: the first covers how the between-subject variables affect decision-making system use, the second looks at how the between-subject variables affect agent performance and the third discusses the results and offers explanations for the observations noted.

Decision-Making System Use. The average total amounts of problem-solving and pattern-recognition system use are shown in Fig. 7. The stacked segments in the figure indicate the proportion of the total taken up by each of the two kinds of decisions: in all cases (for agents of type 2 and 3), the amount of pattern-recognition was significantly smaller than the amount of problem solving.

The results indicated a main effect of intrinsic complexity, $F(2, 243) = 295.2$, extrinsic complexity, $F(2, 243) = 24.0$, and agent type, $F(2, 243) = 1,511.7$ with all $p < 0.001$. As expected, type 1 agents never used pattern-recognition whilst type 2 agents used pattern recognition more frequently than agents of type 3. By increasing extrinsic complexity, agents of type 2 and 3 decreased their use of pattern-recognition; in contrast, increasing intrinsic complexity had the reverse effect. The following interactions were also statistically significant for agent types 2 and 3: intrinsic complexity and agent type $F(4, 243) = 113.6$, $p < 0.001$, extrinsic complexity and agent type: $F(4, 243) = 15.0$, $p < 0.001$.

Average frequency of problem-solving use yielded results that were the mirror-image of those obtained for average frequency of pattern-recognition use. This is expected since, if an agent does not use pattern-recognition then it will use problem-solving. There was a main effect of agent type, $F(2, 243) = 1,512.1$, $p < 0.001$, with type 1 agents using problem-solving most frequently on average and type 2 agents least (as expected). A main effect of intrinsic complexity, $F(2, 243) = 295.2$, $p < 0.001$, reflects that average frequency of problem-solving use tended to decrease with increasing intrinsic complexity. A main effect of extrinsic complexity was significant, $F(2, 243) = 24.0$, $p < 0.001$, reflecting a small increase in problem-solving use as extrinsic complexity increased.

Performance. Figures 8, 9 and 10 show average scores achieved by each agent type for each degree of environment complexity organised by number of

---

[6] Due to an error in the simulation code used in a previous version of this paper [27], the results reported in this section consistently differ from those reported in the corresponding section of [27] by a factor of 10. The results reported in this section use a rectified version of the simulation code and are correct.

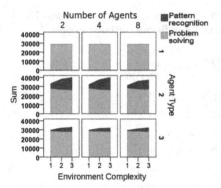

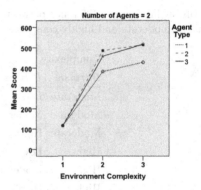

Fig. 7. Average frequency of problem-solving and pattern-recognition system use by agents as a function of agent type and each intrinsic/extrinsic complexity setting outlined in Sect. 5.1.

Fig. 8. Average performance of agents for each intrinsic complexity setting outlined in Sect. 5.1 when 2 agents are present in Tileworld.

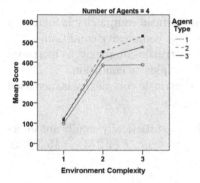

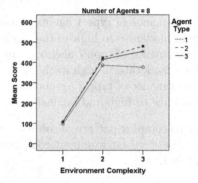

Fig. 9. Average performance of agents for each intrinsic complexity setting outlined in Sect. 5.1 when 4 agents are present in Tileworld.

Fig. 10. Average performance of agents for each intrinsic complexity setting outlined in Sect. 5.1 when 8 agents are present in Tileworld.

agents. The three main effects were statistically significant: intrinsic complexity, $F(2, 243) = 2,437.7$, extrinsic complexity, $F(2, 243) = 16.6$, and agent type, $F(2, 243) = 70.8$ with all $p < 0.001$.

Irrespective of environment complexity, the average score achieved by agents of type 2 was either approximately equal to, or greater than, the average score of type 3 agents. Type 1 agents consistently achieved the lowest average scores. Neglecting average scores achieved when intrinsic environment complexity is low (since there is not much difference between the agent types), agents of type 2 offered a performance increase of up to 36.5 % when compared to agents of type 1 (see Table 3).

Table 3. Percentage difference between average scores achieved by agent types 1 and 2 in the moderate and highly complex environment conditions outlined in Sect. 5.1.

Complexity		Type		
Intrinsic	*Extrinsic*	*1*	*2*	% Diff.
Moderate	Low	381.5	485.0	27.1
Moderate	Moderate	384.3	451.3	17.4
Moderate	High	384.8	421.3	9.5
High	Low	427.5	513.5	20.1
High	Moderate	387.5	529.0	36.5
High	High	375.6	479.4	27.6

By increasing extrinsic complexity, average scores were decreased for each agent type whilst increasing intrinsic complexity caused average scores to increase for each agent type. Exceptions to this trend were:

- Performance of type 1 agents plateaus when intrinsic complexity is increased from moderate to high in the moderate extrinsic complexity condition.
- Performance of type 1 agents decreases when intrinsic complexity is increased from moderate to high in the high extrinsic complexity condition.
- Performance of type 2 agents decreases when extrinsic complexity is increased from low to high and intrinsic complexity is high.

This complex pattern of results is reflected by a statistically significant interaction between environment complexity and agent type, $F(4, 243) = 15.3$ with $p < 0.001$.

Discussion

Decision-Making System Use. We will first consider the effects of increasing extrinsic and intrinsic complexity on decision-making system use. By increasing extrinsic complexity, competition for environmental resources increases. Thus, there are fewer interactions between agents and environmental resources, so fewer productions are created (productions are only ever created when an action is performed successfully, see Sect. 4.3), pattern-recognition system use decreases (since the system does not have the resources required to operate) and problem-solving system use increases. By increasing intrinsic complexity, availability of environmental resources increases. This results in agents interacting with environmental resources more frequently, in turn increasing the number of productions created in LTM. Consequently, the opposite effect to increasing extrinsic environmental complexity is observed: pattern-recognition system use increases and problem-solving system use decreases.

Conversely, *a priori* reasoning would suggest that, when intrinsic environment complexity is increased, the state space of the environment increases: this should result in agents discriminating/familiarising more often, inhibiting the creation of productions (see Sect. 3.1: *Discrimination, Familiarisation and Production Creation.*). Consequently, pattern-recognition use should decrease due to the unavailability of productions. However, when intrinsic complexity is high, it is likely that an agent will encounter similar environment states frequently due to an over-abundance of environmental resources [29]. This, coupled with the fact that the observable space of the environment is relatively small for an agent compared to the total size of the Tileworld environment (25 squares against 1225), means that the number of completely familiarised visual and action chunks in an agent's LTM will increase, facilitating production creation. Thus, it is more likely for an agent that can use pattern-recognition to do so.

Explaining the effect of agent type on decision-making system use is trivial: type 1 agents only use problem-solving, type 2 agents use their pattern-recognition system more frequently than type 3 agents and type 3 agents use problem-solving more frequently than type 2 agents. For agent types 2 and 3: when a production is created for a visual state, type 2 agents will never use their problem-solving system again when that visual state is encountered whereas type 3 agents may do.

Performance. Since increasing extrinsic complexity elevates competition for resources, this should result in performance declining as extrinsic complexity increases; this is observed in most of the data acquired (see Figs. 8, 9 and 10). Increasing intrinsic complexity has the opposite effect since there are more resources available to an agent to achieve their primary goal. Average scores are likely to increase as intrinsic complexity increases. Again, this is observed in most of the data acquired.

To explain the effect of agent type on performance, we must consider the length of time taken to produce an action using either decision-making system against the regularity by which the environment's state may change since this affects whether intention-reconsideration is triggered (see Sect. 4.3). If intention reconsideration is triggered an agent will not perform an action potentially resulting in the relevant environment resources expiring before the agent can use them to achieve its goal.

Since the interval of time for an agent generating an action using problem-solving and the environment potentially creating new tiles and holes is equal (1 s), an agent's intention reconsideration is more likely to be triggered when the problem-solving system is used and intrinsic environment complexity is increased (more so when extrinsic environment complexity is increased too). Therefore, in the space of time where the environment remains static, agents that use pattern-recognition can perform up to 5 actions. This enables agents that employ pattern-recognition more to achieve their goals more quickly and score more frequently. This explains why type 2 agents consistently perform better compared to agents of type 1 and 3 who use pattern-recognition less frequently.

Conclusions. All null hypotheses stated in Sect. 5.1 are refuted. Crucially, the results reported indicate that increased use of the pattern-recognition system rather than the problem-solving system benefits performance since more actions can be performed before the environment state changes due to intrinsic environmental factors (extrinsic factors may still cause state changes, however). This results in an agent's intention reconsideration strategy being triggered less frequently so the agent acts more frequently to achieve its goals than it does deliberating about how to achieve them. This supports the position that acting quickly and, potentially, sub-optimally in complex, stochastic environments benefits performance more than taking time to re-evaluate productions to potentially optimise them. In this sense, the Einstellung effect appears to be beneficial for agents in the environment modelled. This conclusion is further bolstered by the seemingly anomalous effects of increasing extrinsic and intrinsic environment complexity on the performance of type 1 and 2 agents in some conditions (see Sect. 6.1: *Performance*).

Performance for type 2 agents improves when intrinsic complexity is high and extrinsic complexity increases from low to moderate. As explained in the *Decision-Making System Use* and *Performance* discussions above, increasing extrinsic complexity generally impairs performance whilst increasing intrinsic complexity improves performance. The simulation condition outlined therefore appears to strike a balance between these interactions so that the state-space is "just-right" to optimise pattern-recognition use for type 2 agents:

- Agents are not overloaded with visual information causing excessive discrimination/familiarisation and blocking production creation.
- Availability of environmental resources is such that the agent is not blocked from moving/pushing tiles (due to too many resources) and the agent does not spend most of its time looking for resources (due to too few resources).

The same result is not observed for type 3 agents because use of their pattern-recognition system can entail use of problem-solving.

Anomalous results for the performance of type 1 agents can be explained thus. As intrinsic complexity is increased, intention reconsideration will occur more frequently (as already argued). Since type 1 agents can only use problem-solving, their intention-reconsideration will be triggered frequently so they will spend less time acting and more time deliberating, resulting in their performance being hampered.

6.2   Sensitivity Analysis

All results in this section were analysed using a $3 \times 5 \times 9$ ANOVA, with environment complexity, episodic memory size and discount rate as between-subject variables, respectively. We focus on the following dependent variables: average score, average frequency of problem-solving system use and average frequency of pattern-recognition system use. The section is split into three parts: the first presents results concerning decision-making system use, the second presents results concerning performance and the third discusses the results and offers explanations for the observations noted.

Decision-Making System Use. Figures 11 to 14 display results relevant to this section. Note that we do not display results pertaining to the effects of discount rate on decision-making system use since the effect of this variable was, at best, marginal, and, at worst, non-significant. To calculate the results reported, an average of averages for problem-solving/pattern-recognition system-use was calculated. For example, average problem-solving system use reported for type 2 agents with episodic memory size 6 in the low complexity condition was calculated as follows:

1. Calculate the average frequency of problem-solving system use achieved by type 2 agents with episodic memory size 6 and discount rate 0.1 over the 10 repeats for the low environment complexity condition. Repeat for each discount rate.
2. Average the average frequencies from step 1; report result.

Agent Type 2. For average frequency of pattern-recognition system use, there was a main effect of environment complexity, $F(2, 1215) = 2018.0$, $p < 0.001$ and episodic memory size, $F(4, 1215) = 82.8$, $p < 0.001$, but no main effect of discount rate, $F(8, 1215) = 0.5$, $p = $ ns. Only the interaction between environment complexity and episodic memory size was statistically significant, $F(8, 1215) = 2.1$, $p < 0.05$.

Average frequency of problem-solving system use results were a mirror-image to those for pattern-recognition system use. There was a main effect of environment complexity, $F(2, 1215) = 2017.9$, $p < 0.001$ and episodic memory size, $F(4, 1215) = 82.8$, $p < 0.001$, but no main effect of discount rate, $F(8, 1215) = 0.5$, $p = $ ns. The only statistically significant interaction was that between environment complexity and episodic memory size, $F(8, 1215) = 2.1$, $p < 0.05$.

Agent Type 3. For average frequency of pattern-recognition system use, there was a main effect of environment complexity, $F(2, 1215) = 315.1$, $p < 0.001$ and episodic memory size, $F(4, 1215) = 68.1$, $p < 0.001$. Unlike type 2 agents, the main effect of discount rate was marginally significant, $F(8, 1215) = 1.9$, $p = 0.054$ as was the interaction between environment complexity and discount rate $F(16, 1215) = 1.6$, $p = 0.054$; no other interactions were present. While there was still less pattern-recognition system use on average in the high complexity condition (like type 2 agents), there was no significant difference in average frequency of pattern-recognition system use between low and moderate environment complexity conditions, $F(1, 810) = 0.3$, $p = $ ns. This is an important difference between agents of type 2 and 3.

Results for average frequency of problem-solving system use were, again, a mirror image to those obtained for pattern-recognition system use.[7] There was a main effect of complexity and episodic memory size, a marginal effect

---

[7] All F and p values for the effects discussed are equal to those outlined for average frequency of pattern-recognition system use.

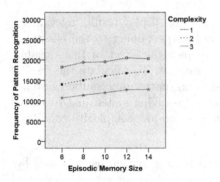

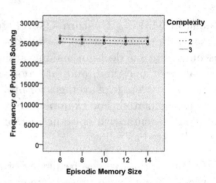

Fig. 11. Effect of episodic memory size upon the average frequency of pattern-recognition system use by type 2 agents for each environment complexity setting outlined in Sect. 5.2.

Fig. 12. Effect of episodic memory size upon the average frequency of problem-solving system use by type 2 agents for each environment complexity setting outlined in Sect. 5.2.

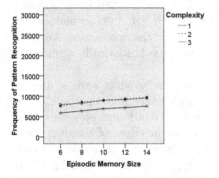

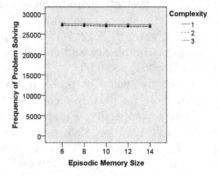

Fig. 13. Effect of episodic memory size upon the average frequency of pattern-recognition system use by type 3 agents for each environment complexity setting outlined in Sect. 5.2.

Fig. 14. Effect of episodic memory size upon the average frequency of problem-solving system use by type 3 agents for each environment complexity setting outlined in Sect. 5.2.

of discount rate and an interaction between environment complexity and discount rate. Again, there is no significant difference in the average frequency of problem-solving system use between low and moderate environment complexity conditions (Fig. 12).

Performance. Figures 15 and 16 display results relevant to this section. As with results concerning frequency of decision-making system use, we do not show results regarding the effects of discount rate upon performance since the effect of this variable was non-significant for both agent types. Results for this section were calculated in the same fashion as results for frequency of decision-making system use (see Sect. 6.2: *Decision-Making System Use*).

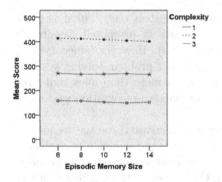

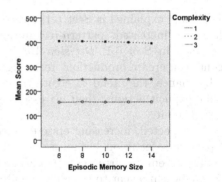

Fig. 15. Effect of episodic memory size upon the performance of type 2 agents for each environment complexity setting outlined in Sect. 5.2.

Fig. 16. Effect of episodic memory size upon the performance of type 3 agents for each environment complexity setting outlined in Sect. 5.2.

Agent Type 2. There was a main effect of environment complexity, $F(2, 1215) = 29,590.9$, $p < 0.001$ and episodic memory size, $F(4, 1215) = 11.8$, $p < 0.001$, but no main effect of discount rate, $F(8, 1215) = 0.521$, $p = $ ns. There was also an interaction between environment complexity and episodic memory size, $F(8, 1215) = 671.6$, $p < 0.01$. No other interactions were statistically significant.

Agent Type 3. Like the performance of type 2 agents, there was a main effect of environment complexity, $F(2, 1215) = 36,627.9$, $p < 0.001$ and no main effect of discount rate, $F(8, 1215) = 1.4$, $p = $ ns. However, the main effect of episodic memory size just failed to reach statistical significance, $F(4, 1215) = 2.1$, $p = 0.074$. There was also an interaction between complexity and episodic memory size, $F(8, 1215) = 3.0$, $p < 0.005$. No other interactions reached statistical significance.

Discussion

Decision-Making System Use. In Sect. 6.1: *Decision-Making System Use*, we noted that increasing intrinsic complexity increased pattern-recognition system use, but increasing extrinsic complexity had the opposite effect. It therefore follows that extrinsic environment complexity must have a more significant effect upon decision-making than intrinsic environment complexity. This is because intrinsic environment complexity increases along with extrinsic complexity in the complexity conditions used in this sensitivity analysis. However, frequency of pattern-recognition use decreases as the conflated in/extrinsic complexity increases.

The observation that frequency of pattern-recognition system use significantly differed for type 2 agents, but not for type 3 agents, when environment complexity was increased from low to moderate must be due to type 3 agents reinforcing problem-solving productions more on average than explicit action productions. To explain: in the low complexity condition, environmental resource availability is low, so problem-solving is used more frequently (as already explained in Sect. 6.1: *Decision-Making System Use*). Then, as complexity is increased, pattern-recognition system use is promoted (again, as explained in Sect. 6.1: *Decision-Making System Use*). However, since type 3 agents can create productions resulting in use of their problem-solving system, if, by chance, they tend to create these productions more than ones that prescribe explicit actions, then increased use of the pattern-recognition system is cancelled out.

As expected, increasing episodic memory size increased average frequency of pattern-recognition system use and decreased average frequency of problem-solving system use, irrespective of agent type. Since increasing episodic memory size enables an agent to store more episodes, this increases the likelihood of the pattern-recognition system being employed since a greater range of visual states will be encoded as productions.

The respective effects of environment complexity and episodic memory size for type 2 agents should be noted: as environment complexity increased, average frequency of pattern-recognition system use slightly decreased. In contrast, as episodic memory size increased, average frequency of pattern-recognition system use slightly increased. Thus, environment complexity and episodic memory size appear to offset one another. However, the F values obtained indicate that environment complexity exerted a more significant effect upon decision-making system frequency than episodic memory size.

Performance. Irrespective of agent type, the best performance was found in the moderate complexity condition, with the worst performance achieved in the low complexity condition. Again, this is most probably because the balance of intrinsic and extrinsic environment complexity is optimal for performance promotion in the moderate complexity condition used in this sensitivity analysis:

- The environment does not change so much that agents spend more time deliberating about how to achieve goals than acting to achieve them, or constantly discriminating/familiarising rather than creating productions so pattern-recognition can be used.
- Intrinsic complexity is not so low that resources are scarce and primary goal achievement is impeded. Neither is it so high that resources can not be moved due to a lack of empty squares.
- Extrinsic complexity is not so high that it is difficult to secure resources required to achieve an agent's goals.

Interestingly, this result differs from the performance of these agent types in their equivalent complexity conditions in the results discussed in Sect. 6.1 (see Figs. 8, 9 and 10). In those original simulations, the performance of type 2 and 3 agents in the moderate complexity condition (value 2 on the x-axis of Fig. 9) is worse than their performance in the high complexity condition (value 3 on the x-axis of Fig. 10). Currently, this discrepancy is unexplained, and will be looked at in future work.

Importantly, smaller episodic memory sizes seem to produce better performance irrespective of environment complexity for type 2 agents (see Fig. 15).

Table 4. Average scores and percentage increases thereof as episodic memory size is decreased from original value specified in simulations run in Sect. 5.1 for type 2 agents across all sensitivity analysis complexity conditions.

Environment complexity	Episodic memory size	Avg. score (% Inc.)
Low	10	152.99
	6	158.52 (3.61)
Mod.	10	409.15
	6	414.30 (1.26)
High	10	267.70
	6	270.77 (1.15)

While this trend also appears to hold for type 3 agents in Fig. 16, its effect was not significant; type 3 agents use their pattern-recognition systems less frequently, on average, than type 2 agents (discussed in Sect. 6.1: *Decision-Making System Use* and reinforced by Fig. 13). By reducing episodic memory size from the median value of 10 (the value used for this variable in the first set of simulations), it is possible to tease out a performance increase of up to 3.61 % (see Table 4).

By retaining fewer episodes, actions that contributed little to the achievement of the agent's primary goal are ignored. This would result in agents creating and reinforcing more useful productions resulting in optimised goal achievement. For type 3 agents it may be the case that productions which prescribe use of the problem-solving system are created and subsequently reinforced more than productions prescribing explicit actions, frequently resulting in protracted periods of deliberation. This would cause either a decrease in performance or, at the least, performance to plateau; the latter was observed for the performance of these agents in low and moderate complexity conditions (see Fig. 16).

As noted in Sect. 6.2: *Performance*, an interaction between environment complexity and episodic memory size occurs. This produced a negative effect on performance: as environment complexity increased, larger episodic memory sizes impinged performance. Since smaller episodic memory sizes reduce the chance of non-optimal productions being used (as argued above) and more complex environments increase the likelihood of an agent reconsidering its intentions and remaining inanimate, it follows that being more discriminating with regard to the utility of one's productions will enhance performance and vice-versa.

Interestingly, this performance impingement was less significant for type 3 agents than type 2 agents, but was less likely to occur due to chance with type 3 agents. This result is most likely produced due to type 3 agents spending more time deliberating than acting, reducing the total number of productions generated. So, whilst type 2 and 3 agents may produce productions whose utility ratings are equivalent, agents of type 3 produce less of them in the same space of time and thus use fewer of them. This accounts for the plateau in performance observed in the low and moderate complexity conditions for type 3 agents too:

agents spend less time reconsidering their intentions due to reduced environment dynamism but the amount of productions executed then attracts a premium. In such circumstances, type 2 agents profit since they can execute more (potentially non-optimal) productions than type 3 agents.

Conclusions. Around half of the null hypotheses stated in Sect. 5.2 are strongly refuted. Varying episodic memory size had a significant effect upon performance and average frequency of problem-solving and pattern-recognition system use for agent types 2 and 3. Reducing episodic memory size slightly improves performance for both types of agent, although no major effects are observed. In contrast, decision-making system use is notably affected by varying episodic memory size: larger memories promote pattern-recognition system use whereas smaller memories promote problem-solving system use. This is important since it was argued in Sect. 6.1 that increased use of the pattern-recognition system benefits performance more so than increased use of the problem-solving system. So, whilst reducing episodic memory size appears to improve performance by creating more useful productions, it also reduces use of the pattern-recognition system in general; a balance must be struck.

Environment complexity also affected performance and decision-making system use significantly: for both agent types, moderate environment complexity produced the best performance whilst low complexity produced the worst performance. With regard to decision-making system use, the effect of environment complexity was significant on both performance and decision-making system use for type 2 agents whereas for type 3 agents, the effect was only consistently significant for performance. The effect of environment complexity on performance is different than that observed in the original simulations and it is our intention to investigate this further in future work.

Altering discount rates did not produce any significant effects upon performance or pattern-recognition system use apart from a marginally significant effect that is observed upon average frequency of problem-solving and pattern-recognition system use for type 3 agents. The reason for this is unclear but given that the effect is hardly significant, an explanation seems unwarranted.

7   Conclusions and Future Work

In this paper we have described and implemented a novel, modular dual-process [26] architecture for self-learning, computational agents. The architecture consists of a problem-solving and pattern-recognition decision-making system, created using a combination of the CHREST architecture, the PSDR algorithm and the Roulette selection mechanism. The system implemented different balances of problem-solving and pattern-recognition use: pure problem-solving, "pure" pattern-recognition and a mixture of both. These balances of the two systems were embodied as three types of agent situated in the Tileworld environment. We used these agents to ascertain how different balances of problem-solving and pattern-recognition system affected performance in this environment given different degrees of intrinsic and extrinsic environmental complexity. We also explored how environment complexity affects agent performance and decision-making before conducting a sensitivity analysis to ascertain if the dependent variables studied in the first set of simulations were significantly altered when salient variables governing the mechanism of the pattern-recognition system were varied.

Use of pattern-recognition was beneficial to agent performance, especially when intrinsic and extrinsic environment complexity was increased, whereas use of problem-solving was less beneficial, due to the required time to solve problems. As overall environmental complexity increased, we found that agents using pure problem-solving (that is, the complete absence of pattern-recognition) are further disadvantaged whereas agents that were more likely to use pattern-recognition performed best. Our results therefore demonstrate that an agent which can use both problem-solving and pattern-recognition is at an advantage in the complex, dynamic environment modelled and even more so when pattern-recognition is favoured. Essentially, the results indicate that agent performance is maximised by generating (potentially sub-optimal) productions quickly and executing many of them, at least in the environment modelled here. This is an interesting finding given that the Einstellung effect [30] is likely to be manifest in the agents that perform best.

The sensitivity analysis performed corroborated these findings, but also revealed the variables which maximise the performance of agents capable of pattern-recognition given different complexities of the environment modelled. We discovered that, whilst episodic memory size affected agent performance significantly (albeit minimally), discount rate did not. Indeed, episodic memory size appears to exert an influence on the balance of agent performance and promotion of pattern-recognition use. Smaller episodic memory sizes enhanced agent performance since only useful productions are created but made it less likely for the agent to actually use pattern-recognition. This is because smaller episodic memory sizes provide the agent with fewer productions, resulting in problem-solving being used instead.

In future work, we would like to ascertain why the effect of environment complexity differs with respect to performance between the two sets of simulations run: is this due to chance or a more exact reason? We also intend to determine if these conclusions still hold when the same simulations are run for longer periods of time, and when heterogeneous agent types compete. Ascertaining to what degree the Einstellung effect is manifest in pattern-recognisers in these simulations would also be a notable addition. Finally, we will consider if these conclusions generalise to other domains and when the amount of information capable of being reasoned with by agents is increased, by expanding the size of their observable environment.

References

1. Aarts, H., Dijksterhuis, A.: Habit as knowledge structures: automaticity in goal-directed behavior. J. Pers. Soc. Psychol. **78**(1), 53–63 (2000)

2. Anderson, J.R., Bothell, D., Byrne, M.D., Douglass, S., Lebière, C., Qin, Y.L.: An integrated theory of the mind. Psychol. Rev. **111**(4), 1036–1060 (2004)
3. Arai, S., Sycara, K.: Effective learning approach for planning and scheduling in multi-agent domain. In: Meyer, J.A., Berthoz, A., Floreano, D., Roitblat, H., Wilson, S.W. (eds.) From Animals to Animats 6: Proceedings of the 6th International Conference on Simulation of Adaptive Behavior, pp. 507–516. MIT Press (2000)
4. Arai, S., Sycara, K.P., Payne, T.R.: Experience-based reinforcement learning to acquire effective behavior in a multi-agent domain. In: Proceedings of the 6th Pacific Rim International Conference on Artificial Intelligence, pp. 125–135 (2000)
5. Baker, J.E.: Reducing bias and inefficiency in the selection algorithm. In: Grefenstette, J.J. (ed.) Proceedings of the 2nd International Conference on Genetic Algorithms on Genetic Algorithms and Their Application. L. Erlbaum Associates Inc. (1987)
6. Bilalić, M., McLeod, P., Gobet, F.: Inflexibility of experts - reality or myth? Quantifying the Einstellung effect in chess masters. Cogn. Psychol. **56**(2), 73–102 (2008)
7. Bossomaier, T., Traish, J., Gobet, F., Lane, P.C.R.: Neuro-cognitive model of move location in the game of Go. In: Proceedings of the 2012 International Joint Conference on Neural Networks (2012)
8. Chase, W.G., Simon, H.A.: Perception in chess. Cogn. Psychol. **4**, 55–81 (1973)
9. Dayan, P., Daw, N.D.: Decision theory, reinforcement learning, and the brain. Cogn. Affect. Behav. Neurosci. **8**(4), 429–453 (2008)
10. de Groot, A.D.: Thought and Choice in Chess (First edition in 1946). Mouton, The Hague (1978)
11. de Groot, A.D., Gobet, F.: Perception and Memory in Chess: Heuristics of the Professional Eye. Van Gorcum, Assen (1996)
12. Erev, I., Roth, A.E.: Predicting how people play games: reinforcement learning in experimental games with unique, mixed strategy equilibria. Am. Econ. Rev. **88**(4), 848–881 (1998)
13. Evans, J.S.B.T.: Dual-processing accounts of reasoning, judgment and social cognition. Annu. Rev. Psychol. **59**, 255–278 (2008)
14. Freudenthal, D., Pine, J.M., Gobet, F.: Simulating the referential properties of Dutch, German and English root infinitives in MOSAIC. Lang. Learn. Dev. **15**, 1–29 (2009)
15. Gillan, C.M., Papmeyer, M., Morein-Zamir, S., Sahakian, B.J., Fineberg, N.A., Robbins, T.W., de Wit, S.: Disruption in the balance between goal-directed behavior and habit learning in obsessive-compulsive disorder. Am. J. Psychiatry **168**, 718–726 (2011)
16. Gobet, F.: Les mémoires d'un joueur d'échecs. Editions Universitaires, Fribourg, Switzerland (1993)
17. Gobet, F.: A pattern-recognition theory of search in expert problem solving. Thinking Reasoning **3**, 291–313 (1997)
18. Gobet, F., Lane, P.C.R., Croker, S.J., Cheng, P.C.H., Jones, G., Oliver, I., Pine, J.M.: Chunking mechanisms in human learning. Trends Cogn. Sci. **5**, 236–243 (2001)
19. Grefenstette, J.J.: Credit assignment in rule discovery systems based on genetic algorithms. Mach. Learn. **3**, 225–245 (1988)
20. Hesketh, B.: Dilemmas in training for transfer and retention. Appl. Psychol. **46**(4), 317–339 (1997)
21. Holroyd, C.B., Coles, M.G.: The neural basis of human error processing: reinforcement learning, dopamine, and the error-related negativity. Psychol. Rev. **109**(4), 679–709 (2002)

22. Jones, G.A., Gobet, F., Pine, J.M.: Linking working memory and long-term memory: a computational model of the learning of new words. Dev. Sci. **10**, 853–873 (2007)
23. Jongman, R.W.: Het Oog Van De Meester. Van Gorcum, Assen (1968)
24. Kheirbek, M.A., Klemenhagen, K.C., Sahay, A., Hen, R.: Neurogenesis and generalization: a new approach to stratify and treat anxiety disorders. Nat. Neurosci. **15**(12), 1613–1620 (2012)
25. Laird, J.E.: The Soar Cognitive Architecture. MIT Press, Cambridge (2012)
26. Lane, P.C.R., Gobet, F.: CHREST models of implicit learning and board game interpetation. In: Bach, J., Goertzel, B., Ikle, M. (eds.) Proceedings of the 5th Conference on Artificial General Intelligence. LNAI, vol. 7716, pp. 148–157. Springer, Heidelberg (2012)
27. Lloyd-Kelly, M., Gobet, F., Lane, P.C.R.: The art of balance: problem-solving vs. pattern-recognition. In: Proceedings of the 7th International Conference on Agents and Artificial Intelligence, pp. 131–142 (2015)
28. Lloyd-Kelly, M., Gobet, F., Lane, P.C.R.: Piece of mind: long-term memory structure in ACT-R and CHREST. In: Noelle, D.C., Dale, R., Warlaumont, A.S., Yoshimi, J., Matlock, T., Jennings, C.D., Maglio, P.P. (eds.) Proceedings of the 37th Annual Meeting of the Cognitive Science Society, pp. 1422–1427. Cognitive Science Society (2015)
29. Lloyd-Kelly, M., Lane, P.C.R., Gobet, F.: The effects of bounding rationality on the performance and learning of CHREST agents in Tileworld. In: Bramer, M., Petridis, M. (eds.) Research and Development in Intelligent Systems XXXI, pp. 149–162. Springer International Publishing, Switzerland (2014)
30. Luchins, A.S.: Mechanization in problem solving: the effect of Einstellung. Psychol. Monogr. **54**(6), 1–95 (1942)
31. Miller, G.A.: The magical number seven, plus or minus two: some limits on our capacity for processing information. Psychol. Rev. **63**, 81–97 (1956)
32. Miyazaki, K., Yamamura, M., Kobayashi, S.: On the rationality of profit sharing in reinforcement learning. In: 3rd International Conference on Fuzzy Logic, Neural Nets and Soft Computing, pp. 285–288. Korean Institute of Intelligent Systems (1994)
33. Pollack, M., Ringuette, M.: Introducing the Tileworld: experimentally evaluating agent architectures. In: 8th National Conference on Artificial Intelligence, pp. 183–189. AAAI Press (1990)
34. Raza, M., Sastry, V.: Variability in behavior of command agents with human-like decision making strategies. In: 10th International Conference on Computer Modelling and Simulation, pp. 562–567 (2008)
35. Saariluoma, P.: Error in chess: the apperception-restructuring view. Psychol. Res. **54**, 17–26 (1992)
36. Samsonovich, A.: Toward a unified catalog of implemented cognitive architectures. In: Proceedings of the 2010 Conference on Biologically Inspired Cognitive Architectures, pp. 195–244. IOS Press, Amsterdam, The Netherlands (2010)
37. Simari, G.I., Parsons, S.D.: On approximating the best decision for an autonomous agent. In: 6th Workshop on Game Theoretic and Decision Theoretic Agents, pp. 91–100. Third Conference on Autonomous Agents and Multi-agent Systems (2004)
38. Simon, H.A.: The Sciences of the Artificial. MIT Press, Cambridge (1969)
39. Simonton, D.K.: Origins of Genius: Darwinian Perspectives on Creativity. Oxford University Press, New York (1999)
40. Sloman, S.: The empirical case for two systems of reasoning. Psychol. Bull. **119**, 3–22 (1996)

41. Sternberg, R.J.: Costs of expertise. In: The Road to Excellence: The Acquisition of Expert Performance in the Arts and Sciences, Sports, and Games, pp. 347–354. Lawrence Erlbaum Associates, Hillsdale, NJ (1996)
42. Sun, R., Merrill, E., Peterson, T.: From implicit skills to explicit knowledge: a bottom-up model of skill learning. Cogn. Sci. **25**, 203–244 (2001)
43. Sun, R., Slusarz, P., Terry, C.: The interaction of the explicit and the implicit in skill learning: a dual-process approach. Psychol. Rev. **112**(1), 159–192 (2005)
44. Sutton, R.S., Barto, A.G.: Reinforcement Learning: An Introduction. MIT Press, Cambridge (1998)
45. Watkins, C.J.C.H., Dayan, P.: Technical note: Q-learning. Mach. Learn. **8**, 279–292 (1992)
46. de Wit, S., Dickinson, A.: Associative theories of goal-directed behaviour: a case for animal-human translational models. Psychol. Res. **73**(4), 463–476 (2009)
47. Zeitz, C.M.: Some concrete advantages of abstraction: how experts' representations facilitate reasoning. In: Expertise in Context: Human and Machine, pp. 43–65. The MIT Press, Cambridge, MA (1997)

Author Index

Printed in the United States
By Bookmasters